Praise for *Mr. America*

"This book is a delightful and colorful piece of true Americana. Surely other countries must also sometimes give birth to figures like Bernarr Macfadden—a brilliant, ambitious, competitive, compulsively muscular, early twentieth-century multimillion-dollar marketeering genius who launched a wildly successful (though peculiar) health and diet fad. But we Americans really always have done that kind of thing better than anyone else—as this great big story proves. Enjoy every twist and turn of it."

—Elizabeth Gilbert, author of *Eat, Pray, Love*

"Impressive.... Adams chronicles how, in the first half of the twentieth century, Macfadden opened some of the country's original health food restaurants and communes, assembled early bodybuilding shows, and built a $30 million magazine empire.... Adams provides historical context throughout and tells the story with journalistic remove, resisting much commentary, even as the reader is busy filling the margins with exclamation points."

—*The New York Times Book Review*

"Every decade ushers in a new 'revolutionary' exercise fad (or three). But as *Mr. America* points out, most of them probably owe a nod to Bernarr Macfadden and his magazine *Physical Culture*."

—*Men's Journal*

"Imagine if Rupert Murdoch, Jack LaLanne, and Dr. Andrew Weil all got together and had a baby, then raised that child on wheat germ and 100 pushups a day. Only such a prodigy could give you a sense of the sheer eccentric magnificence of Bernarr Macfadden. Luckily, Mark Adams has brought this lost American legend back to life for us, with surprising insight, sly humor, and even some handy tips on fasting."

—John Hodgman, author of *More Information Than You Require*

"With such a great eccentric as his subject, Mark Adams shrewdly gets out of the way (and gives Macfadden's extreme weight-loss regimens a try)."

—*Details*

"A wild story.... Combining a healthy love for the undraped human body with a brutal dedication to stark privation in perfecting it, Macfadden comes across as one of the twentieth century's most driven and fascinating kooks, fighting to liberate America from the chains of prudery, appetite, the medical establishment, and good taste."

—*Reason*

"When you read this book, you'll be astonished you hadn't heard of Bernarr Macfadden—one of the most fascinating, brilliant, bizarre, and influential characters in American history. I know I was. I want to thank Mark Adams for bringing him to light in this great biography, which you'll devour like a bowl of Macfadden's Strengtho cereal."

—A. J. Jacobs, author of *The Year of Living Biblically*

"A witty, perfectly pitched re-creation of a long-forgotten and madly outsized figure. Adams will make you laugh with fresh information on every page, and in this age of flabby, overlong biography, his delightful production has a leanness that Macfadden himself would approve of."

—Thomas Mallon, author of *Bandbox* and *Dewey Defeats Truman*

"Laugh-out-loud funny.... Would there be such things as the *SI* Swimsuit Issue, the detox fast, and socially acceptable conversations about poop without Bernarr Macfadden? Possibly. But the world is so much more amusing when you imagine it all starting with this strange little man."

—*GQ.com*

"*Mr. America* is required reading for anyone committed to increasing the quality of their life."

—Baron Baptiste, author of *Journey Into Power*

"A delightful read.... It makes no difference whether you start at the beginning of this book, with author Mark Adams's lively story of the bizarre life of America's original health guru, or with the appendix, where Mr. Adams reports on how he himself fared when he put some of Bernarr Macfadden's advice into practice."

—*The Washington Times*

★ M R . ★
AMERICA

★ MR. ★
AMERICA

How Muscular Millionaire Bernarr Macfadden
Transformed the Nation Through
Sex, Salad, and the Ultimate Starvation Diet

MARK ADAMS

itbooks

AN IMPRINT OF HARPERCOLLINS PUBLISHERS

Except as noted below, images are from the author's collection. Grateful acknowledgment for permission to reproduce images is made to the following:

Fulton Oursler, Jr.: page 1, top left, top right, bottom right; page 3, top; page 4, top left; page 5, top right, bottom left, bottom right; page 7, top. *Library of Congress*: page 2, top, middle. *New York Public Library Digital Collection*: page 2, bottom. *Collection of Joe Sapia*: page 3, bottom. *Collection of Professor Jacqueline Reich*: page 7, bottom. *Jane Schryver*: page 8, bottom.

FIRST IT BOOKS EDITION PUBLISHED 2010.

Designed by Renato Stanisic

Library of Congress Cataloging-in-Publication Data

Adams, Mark.
  Mr. America : how muscular millionaire Bernarr Macfadden transformed the nation through sex, salad, and the ultimate starvation diet / Mark Adams.—1st ed.
    p. cm.
  ISBN: 978-0-06-059475-6
  1. Macfadden, Bernarr, 1868–1955. 2. Macfadden, Bernarr, 1868–1955—Influence. 3. Physical education and training—United States—Biography. 4. Health reformers—United States—Biography. 5. Physical fitness—United States—History—20th century. 6. Bodybuilding—United States—History—20th century. 7. Health attitudes—United States—History—20th century. 8. Publishers and publishing—United States—Biography. 9. Periodicals—Publishing—United States—History—20th century. 10. Millionaires—United States—Biography. I. Title.
  CT275.M1364A34 2009
  613.7092—dc22
                                                          2008018705

ISBN 978-0-06-059476-3 (pbk.)

09 10 11 12 13 WBC/RRD 10 9 8 7 6 5 4 3 2 1

For My Father

Contents

The first thing I would do to a friend of mine who happened to be ill with an acute disease would be to put a sign in front of him that he could see every minute: "Food is Poison—Don't Eat if You Want to Keep Out of the Cemetery." That is an absolute statement of facts.

—BERNARR MACFADDEN, *PHYSICAL CULTURE*, JUNE 1916

★ M R . ★
AMERICA

Prologue

At around 10:15 on the muggy evening of Sunday, August 26, 1951, an eighty-three-year-old man in a rumpled suit arrived at the CBS studios on West Forty-seventh Street in New York City, where he was quietly escorted to the control booth of one of television's first runaway hits: *What's My Line?* Thirty million people were about to tune in to the live broadcast, on which a glib quartet of panelists dressed in tuxedos and ball gowns attempted to guess the occupations of a quirky cavalcade of visitors—hatcheck girls, cookbook authors, fire-eaters, elevator operators. The evening's highlight came in the middle of the show, at 10:45. With the four panelists now blindfolded, studio announcer John Daly leaned in to his microphone to ask the question everyone at home was waiting for: "Will you come in, mystery challenger, and sign in, please?"

A summons to serve as the mystery guest on *What's My Line?* was a confirmation of position in the celebrity cosmos; the show's talent bookers were instructed to invite only "public figures who were recognizable on sight by the majority of the television audience." In this most entertaining segment, the famous visitors, attempting to throw off the blindfolded panelists, would ridiculously distort their unmistakable voices so as not to reveal their identities. During the first three weeks of August 1951 the mystery guests had been Yogi Berra, Chico Marx, and Buster Keaton, and like them, on this late-summer evening, the old man with the untidy mop of white hair needed no introduction. An admiring three-part profile that ran in the *New Yorker* the previous year had summed up his rise from penniless orphan to multimillionaire as "one of the great American sagas." His face and name had appeared thousands of times in newspapers and magazines; his nasal Missouri twang was instantly identifiable from his years of broadcasting a morning exercise show on the radio; his trademark sculpted torso had starred in countless newsreels.

If the eighty-three-year-old fidgeted as he sat in the mystery guest's chair with comically correct posture, it was because he was wearing shoes, which he disliked. Nothing thrilled him more than attention, so he was surely delighted to know that when the morning papers landed on doorsteps, his age-defying exploits would be the talk of the nation yet again: Just hours before the TV broadcast, the octogenarian had belatedly celebrated his birthday by jumping two thousand feet from a Stinson monoplane into the Hudson River. Dressed in red flannel underwear, a white life jacket, and a football helmet, he'd landed on his back about twenty-five feet from the New Jersey shore, struggled briefly to untangle himself from his chute, and was collected by a fishing boat. Immediately afterward, he invited the gaggle of reporters assembled on the riverbank back to his Manhattan apartment, not for cake and coffee (neither of which ever passed his lips) but for a lecture on some of his favorite topics: the importance of low-fat eating, power walking, and sunbathing in the nude.

In 1951, Bernarr Macfadden was undeniably one of the most famous men in America. Had he needed to recite his résumé for the *What's My Line?* audience, viewers might have been kept up half the night. His curriculum vitae seemed to cover the careers of half a dozen very different men. Though Macfadden often gave the impression of being functionally illiterate, he'd built one of the largest publishing companies in the country and dreamed up at least one magazine, *True Story*, that was so profoundly new and influential that its legacy is still visible on television twenty-four hours a day. He founded the *New York Evening Graphic*, widely considered to be the worst newspaper in U.S. history, but one that incubated Walter Winchell, Ed Sullivan, and other founders of the celebrity gossip industry. He was the driving force behind the rise of bodybuilding, hand-picking an anonymous Brooklyn model named Charles Atlas as the world's most perfect man, then sketching out a career path that would earn the strongman fame and fortune. Macfadden was a vocal opponent of sexual repression—he'd been sentenced to hard labor for publishing information that demystified the reproductive act while Sigmund Freud was still in Vienna tinkering with his early theories of libido. He once hired Eleanor Roosevelt to edit a magazine about babies; he ran for the Senate on a political platform largely based on the evils of white bread; and he founded a religion that combined Christian doctrine with exercise and diet tips. Perhaps his dearest belief was that almost any disease would disappear if the patient just stopped eating.

The day after his stint on *What's My Line?*—in which panelist Bennett Cerf quickly guessed his identity—papers quoted him as he'd been fished out of the Hudson River: "I feel like a million. I feel 25. Next year, I'm going to jump over Niagara Falls." Instead, he celebrated his eighty-fourth birthday in Paris by parachuting into the Seine. The French papers hailed him as *le grand-père volant*, the flying grandfather. Everyone had a good laugh.

And then Bernarr Macfadden vanished into history.

Like most Americans too young to remember *What's My Line?* I had never heard of Bernarr Macfadden ten years ago, when I was named health editor of *GQ* magazine. The promotion came as something of a surprise, since I knew next to nothing about diet and exercise. I was fairly certain that America's love-hate relationship with fitness somehow wound backward through the neon and spandex step-aerobics era of Richard Simmons and Jane Fonda, past the jogging craze and health clubs and shiny Nautilus machines. Beyond that? Ummm . . . didn't Jack LaLanne invent exercise in the 1950s? I assumed that eating "healthy" foods, like the dense wheat breads that my mother had forced on me as a child of the seventies (under the threat of revocation of *Laverne and Shirley* privileges) was the legacy of Adelle Davis. Bodybuilding, naturally, started with Arnold. America had been a sexual wasteland until the arrival of *Playboy*, according to every Hugh Hefner interview I'd ever read. Preventive medicine and holistic treatments were New Age innovations from my own adulthood, the province of gurus like Deepak Chopra and Dr. Andrew Weil.

Faced with the prospect of acting as a professional health expert, I did what most magazine editors do in such a crisis: I hunted up old magazines to crib ideas from. Fortuitously, I spotted a stack of crumbling issues of *Physical Culture* in a junk shop in Ithaca, New York. I flipped open the topmost copy, dated September 1925, to the *Editor's Viewpoint* column. I might as well have lifted the lid on a music box, for I instantly heard a voice booming at me through the decades: "WEAKNESS IS A CRIME! DON'T BE A CRIMINAL!" The prose was fearless, mesmerizing, and maybe a little unhinged. The table of contents, printed during the heyday of flappers and speakeasies, read like a health bulletin beamed back from a more salubrious future:

Raw Foods Cured My T.B.
Fresh Air + Diet + Exercise = a Good Job
My Fat Is Going Away and I'm Coming Back
The Havoc Wrought by Beauty Doctors
75 Billion Cigarettes a Year Sapping the Nation's Strength

Other issues contained stories about yoga, birth control, weight lifting, homeopathy—and advertisements for an alternative-health wonderland called the Physical Culture Hotel, which had once existed only an hour's drive from where I stood. I carried the entire pile to the cash register, sensing, as many *Physical Culture* readers had long ago, that the magazine was going to change my life. Its editor was Bernarr Macfadden.

If Macfadden has been forgotten—something he would have found utterly inconceivable in August 1951—it is partly his fault. He was notoriously secretive; one employee, a good friend, marveled at his "genius for enveloping himself in a cloud of intrigue." He left no papers from the first seventy years of his life. In terms of business letters and personal correspondence, Pliny the Younger bequeathed his biographers more raw material to work with. But Macfadden devoted his life to his ideas on health, and published tens of thousands of pages of those unfiltered thoughts in his books, newspapers, and magazines, especially *Physical Culture*, which is surely one of the strangest publications of major influence that appeared in the twentieth century. (Close to fifty million copies were sold between the two World Wars.) Every person I showed the magazine to had the same reaction: "I can't believe this thing actually existed."

Macfadden's fellow mystery guest Buster Keaton is remembered as one of the great physical comedians and directors in Hollywood history. Chico Marx is one of the two memorable speaking Marx brothers—the one who wasn't Groucho. Yogi Berra is a member of the Baseball Hall of Fame. And Macfadden? His impact on America was far greater than the other three combined, both for better and for worse. His story is an odd one, sometimes verging on the fantastic, but to the best of my knowledge, everything that follows is true.

Chapter 1

Little Orphan Bernie

Winston Churchill has said that all the famous men of his
acquaintance were the products of an unhappy childhood.
Macfadden cannot remember having had even a moderately
cheerful day before he was twenty.

—ROBERT LEWIS TAYLOR, *NEW YORKER*, OCTOBER 14, 1950

No one in the tiny Ozark town of Mill Springs, Missouri, was likely to have been surprised when William McFadden decided to drink himself blind one day in 1873. McFadden, a Union Army veteran with a menacing face and bushy beard, was a sometime farmer with only three interests in life: hunting deer, drinking corn whiskey, and playing the horses. In between these pursuits he would sometimes endure a few hours with his family, which consisted of his young wife, Mary, and their three children, the newborn baby Alma, three-year-old Mary, and four-year-old Bernard.

Gregarious in his few sober moments, William McFadden was transformed by alcohol into a monster. Years later, his famous son would recall the unhappy cycle of life in the two-room McFadden home. William would begrudgingly earn a few dollars from farm labor, invest his earnings in whiskey and long shots at the racetrack, then return home to the farmhouse on the Black River to beat his family.

What surely *did* surprise the residents of Mill Spring on that day in 1873 was that Mary had decided her marriage was over. She was not a strong woman, and at age twenty-seven she was thirteen years younger than her large, powerful husband. Once already she had run away to her parents' house with her two eldest children to escape William's abuse and give birth to Alma. This time, though, Mary had reason to believe she was leaving for good. She was ill with tuberculosis, a near-certain death sentence.

Mary packed up the children and crossed the Black River in the family's small skiff for the last time, pulling hand over hand on a rope strung across the water. She moved back in with her parents, the Millers, who lived in Ironton, forty miles upstate. A penitent William occasionally sobered up long enough to make the journey north and lurk outside the Millers' log fence, in hopes of rekindling his relationship with his wife. Instead, Mary sued for divorce. Bernard and his sisters never saw their father again. Within a year of their separation, he was dead from delirium tremens.

The Ozarks were a particularly bleak place in the 1870s. Missouri was bogged down in Civil War debt and battered by an economic recession, one of the worst in U.S. history. Missouri had, essentially, endured its own civil war. A large chunk of the state was taken and held by Confederates until 1862, and while Missouri nominally owed its loyalty to the Union, an estimated twenty-five to thirty thousand of its men fought for the Southern cause, including Frank and Jesse James, whose daylight robberies would terrorize the region into the 1880s. Its already poor roads were trampled by soldiers on the march, its draft horses had been conscripted by armies on both sides, and its rich farmland, where countless anonymous battles took place, needed to be reclaimed after years of disuse. Mill Springs was hit by a cholera epidemic the same year as Mary's final escape to Ironton. Two summers later, the state's crops were devastated by a plague of grasshoppers.

Against this backdrop, little Bernard grew into something of a mama's boy, a delicate child whom the local boys took malicious glee in dunking mercilessly in the local rivers. The earliest known photograph of Bernard, taken at around age four, shows a passive child with a weak mouth. He appears to be waiting for someone to take him by the hand and lead him somewhere. Perhaps he'd had a premonition of the childhood odyssey that lay before him.

Five years of suffering began at around age seven, when Bernard, who remembered being ill for most of his childhood, was vaccinated for smallpox in the manner standard at the time—by having a scab from a smallpox lesion applied to a cut in his arm. The price of immunity in his case was six months in bed from blood poisoning.

One morning not long after Bernard had recovered, his mother took him to St. Louis. They were met at the Mississippi River docks by a strange gentleman. Mary explained to her son that the man was going to take him away on a steamship. She did not mention a return trip. Long after the boy had grown up and reinvented himself as Bernarr Macfadden, he recalled the resulting scene as being "torn screaming and clawing and kicking in a frantic agony of fear" from his mother's arms. The man managed to pull Bernard from Mary and lead him toward the wharf, but the boy broke free and ran back to his mother, tears running down his face. Mary told her eight-year-old son the cold truth. Hopeless and nearly destitute, wasting away from late-stage tuberculosis, she no longer had the energy or means to care for a growing boy. She was sending him off to the cheapest boarding school she could find.

Bernard wouldn't learn much at the school, whose name is unknown but which Macfadden later referred to as an "orphan's home" and "the Starvation School." In his opinion, the gruel-fed orphans in *Oliver Twist* were overstuffed gluttons compared to his classmates, one of whom took him aside upon his arrival and whispered, "You'll find out. They never feed us *nothin'*." In truth, the headmaster did feed them something, and that something was peanuts. In the years before George Washington Carver alchemized goobers into everything from soap to axle grease, the legumes were sold as hog feed at about a dollar a ton. If a boy at the Starvation School found himself in possession of a nickel, it was often invested in more peanuts, which would be consumed "shells and all," Macfadden said. Were one of the students blessed with the bounty of an entire apple, "no boy ever asked for the core, for there was no core to give away."

On occasions that the school's directors visited, a full feast would be placed theatrically in front of the wide-eyed pupils. Judging from an editorial Bernarr Macfadden wrote in 1933, though, when he could have purchased enough peanuts to fill the Chrysler Building, those visits must have been infrequent: "If the writer lives a thousand years he will never forget the starvation diet furnished at that school. He was voraciously hungry at all times." When his mother arrived unexpectedly after several months to reclaim him, his most vivid memories were not of their reunion but of the meal they ate.

The reprieve was a short one. With Bernard at school, Mary had managed to place his sisters with relatives and had cooked up a similar—and, she hoped, permanent—solution for the problem of what to do with her only son. Soon after his ninth birthday, he was sent off to work in a dreary, inexpensive hotel in Mount Sterling, Illinois, a town of 1,400 located 250 miles north of Mill Spring. The proprietors were a married couple and distant relatives of Mary's. Macfadden left their names out of his published memoirs, but the wife certainly left a strong impression on the boy. She was strict but good-natured, he wrote, the brains behind the hotel but also—and there was no greater sin to the adult Bernarr Macfadden—"grotesquely fat."

Their familial greeting to their kinsman made clear the level of sentiment he could expect. "He ain't much to look at, but there's likely work in him," said the woman. "If there's work in him, we'll get it out," added her husband.

Bernard was set to task immediately, emptying the chamber pots from guest rooms, blacking shoes, washing laundry, scrubbing floors, and helping drunks to bed. He usually collapsed into his own bed around midnight. Occasionally, to satisfy a truant officer, he would spend a day in school.

In return for working hundred-hour weeks, nine-year-old Bernard received room and board. Any spare moments were spent at the front window, looking longingly down the street toward the train station, waiting for his mother to take him home once more.

One day the proprietor pulled Bernard aside. "Boy, I've got some news for you," he told him. "Your mother's dead."

"And if you ask me, this one's going the same way soon," added the wife.

"He's got all the symptoms," said the husband. "Consumption runs in the family."

Bernard, believing that he'd just heard his own death sentence, ran to bed, pulled the sheets over his head, and sobbed for himself and his dead mother.

As the months passed, Bernard noticed that fewer and fewer customers required his round-the-clock services. The couple eventually quit the hospitality business. With no further use for their junior relative, they turned him over to a farmer who'd expressed interest in adopting a boy. In return for Bernard, he told a reporter much later, they received a small amount of cash and "a scattering of mixed produce."

Today, Macomb, Illinois, is best known as home to Western Illinois University, one of the least inspiring college campuses in the Midwest. There wasn't much to recommend the town in 1879, either, other than its location equidistant from Chicago, St. Louis, and Des Moines. Macomb's newest resident seemed happy to be there, though, and well-prepared for his new life. Though orphaned, penniless ("the only bequest from his parents was what was then popularly known as a delicate constitution," wrote the hammiest of his three authorized biographers), and borderline illiterate, the eleven-year-old had already grasped a cold frontier truth—in an unfriendly world, the only person he could depend on was himself.

As if to test that wisdom, fate delivered him into the hands of Robert Hunter. Farmer Hunter was notoriously cheap; a neighbor once dubbed him "the stingiest son of a bitch east of the Mississippi River." Bernard was again put to work, tending livestock, churning butter, and babysitting Hunter's two young children.

Bernard had already made three discoveries that would inform his health philosophy later in life: the potential downside of vaccination; the human body's ability to survive on a meager diet; and the importance of efficient hard work. The shift from hotel busywork to rigorous farm labor at the outset of his adolescence led him to yet another revelation—that of the rapid physiological changes that could be achieved with pure food and rigorous exercise. On Hunter's farm, he developed a lifelong taste for milk and fresh vegetables—neither of which was yet a staple of the American diet—and was amazed by how churning butter built up his skinny biceps and triceps.

On the occasional days that Bernard made it to school, he found that math came naturally, but he struggled with grammar. He liked to read, though, mostly sentimental stories that made him cry. One morning, after a fire-and-brimstone sermon at church left him feeling that he was damned regardless of his behavior, he slipped away with a plug of tobacco and a jug of Hunter's applejack. Several hours of self-medication did nothing to ease his preadolescent Weltschmerz, but it did accomplish one thing: The ensuing two-day hangover was sufficient to teach him another essential lesson, about the evils of alcohol and tobacco.

Bernard endured his first year at Hunter's farm in reasonably good spirits, and over time he took on more and more duties, to the point where he was milking ten to twelve cows each night. He marveled at his new strength and took immense pride in thrashing a bully who'd mocked his unusual home life. He chafed at his foster father's cheapness, however, when it came to things like providing adequate winter footwear for a region where overnight lows in January average fourteen degrees. He later recalled "shucking corn in freezing temperature, resulting in sore hands and frost-bitten feet." Hunter refused to buy him a new pair of boots, so the boy had his repaired for seventy-five cents, roughly 60 percent of the savings he'd amassed by selling some of Hunter's maple syrup to a neighbor on the sly.

Even before their battle over boots, Bernard and his foster father had grown to dislike each other. The boy resented receiving nothing but room and board for performing the same work as an adult farmhand. Hunter was displeased that his ward had missed several days of work after severely spraining an ankle; he'd been thrown from a horse at the end of an impromptu race against a neighbor boy. Bernard began to consider running away. He took the extra step of discreetly sending a letter to a credit bureau to inquire about the financial health of an uncle in southern Illinois. (Word came back that the man was a horse thief, and not a particularly good one.) After one final quarrel, Bernard waited for a moonless night and, presumably shod in his freshly repaired boots, hiked the two miles to town. There he hopped a train to return to his hotelier relatives in Mount Sterling, who were looking a lot better in retrospect.

Life had changed dramatically for his kin in two years. The husband was dead and the daughters had been married off. Left on her own, the enormous aunt was running a new, smaller hotel by herself. Bernard sensed that she was intimidated by the physical changes in him, and she quickly shipped him off to St. Louis, where his grandmother and two uncles were living. Since he was now penniless, he had to adopt the budget-travel strategy of the day: Board a train, keep a low profile, hope not to get kicked off for being ticketless, get kicked off anyway, walk to the next town, and repeat as necessary. Eventually, he arrived in St. Louis, where he was greeted by the biggest shock in a brief lifetime of traumas: family members who were happy to see him.

Chapter 2

The Great Awakening

There is a serious disinclination among most of our people to take that amount of physical exercise which is necessary to the full enjoyment of all the faculties and the promotion of health. . . . It is sad to think that so many people are overlooking the vitally important fact that physical culture is equally as essential as mental training.

—*BOSTON GLOBE* EDITORIAL, SEPTEMBER 19, 1881

Though he'd been an orphan for four years, Bernard had not completely lost touch with his extended family. Each Christmas he'd received a package from his uncle Harvey Miller in St. Louis, a box containing oranges, nuts, and dried fruits. The thrill of unwrapping such delicacies in the midst of a prairie winter had planted the idea in Bernard's mind that Harvey, with whom his sister Mary had gone to live after their mother's death, was a man "rolling in wealth." Upon his arrival in St. Louis, Bernard learned that his uncle was actually the bookkeeper in a dry-goods store and that Harvey was already supporting his own mother— Bernard's grandmother—as well as Bernard's sister.

Bernard immediately took a job with Dun and Co. (the precursor to the modern Dun and Bradstreet) as a route boy. He earned $12 a month. Having endured savage beatings from his drunken father as a toddler, near-starvation (both physical and emotional) at an orphans' home, two episodes of domestic serfdom, and the death of both parents, Bernard, perhaps not surprisingly, had by the age of twelve developed an unshakable confidence. In an extraordinary photograph taken with Harvey's brother Crume Miller around this time, Bernard wears a top hat, morning coat, and a look of defiant impatience well beyond his years.

After a year of credit reporting, Bernard moved up to a job as office boy at Uncle Harvey's grocery store. The money was an improvement, but long hours at the deskbound job soon sapped his hard-won physical strength. Coughing fits kept him awake through the night. His Uncle Crume was overheard whispering the word "consumption." Severe headaches joined his list of maladies. Doctors were no help. He sampled a variety of patent medicines, those fraudulent remedies whose alleged curing powers usually came from alcohol or opiates, but these had no effect. "I was a complete physical wreck," he remembered.

His weight fell to almost a hundred pounds. "I was at an age just approaching manhood," he later wrote in a style that suggests he hadn't completely abandoned his romance-novel habit, "and I was to be denied health, the only gift in connection therewith which was really of any value."

Almost all self-help icons have creation myths, which turn on a moment when they become weak or debased, hit bottom, then choose to redeem their lives. Theodore Roosevelt's chest-thumping *Autobiography* could have been titled *Up from Asthma*. Oprah Winfrey has written of the dark day she poured syrup over a package of half-frozen hot dog buns and wolfed down the lot. For Bernarr Macfadden, the sickly winter of 1882–83 was the lowest moment of his life.

St. Louis in 1880 was a boomtown, a city whose population had doubled in twenty years and would almost double again in the next twenty. In the decade following the European revolutions of 1848, more than a million Germans flooded into the United States. St. Louis, a hub of rail and river traffic on the eastern edge of the American frontier, was a popular destination for these new arrivals. By 1860, almost a third of the city's 160,000 residents were German-born. (Included in their numbers was a former river pilot from the Rhineland, Adolphus Busch, whose decision to take a job at Eberhard Anheuser's brewery would affect generations of Budweiser drinkers.) These immigrants were fierce about preserving their old-country ways—German was a required subject in St. Louis public schools through 1888. One of those imported traditions would set Bernard on the course that would make him famous.

Around noon one day early in 1883, Uncle Harvey and his nephew were walking in downtown St. Louis and happened to pass the Missouri Gymnasium on the corner of Ninth and St. Charles. This curious structure, carved out of an abandoned church, was plastered with posters of musclemen. Bernard had read newspaper articles about the health benefits of exercise and was intrigued enough to have a look inside. Upon crossing the building's threshold, he felt a tingling sensation. Stepping into the overhead gallery that had once been a choir loft, the fifteen-year-old Bernard looked down on a roomful of sinewy men dressed in attire ranging "from trousers and armless shirts to tights that covered the body only from the waist to foot. They all seemed strong, active, and healthy; even the worst specimen was an Apollo compared to my wasted, angular body."

The he-men who mesmerized young Bernard were German immigrants. Among the cultural traditions they'd imported from the fatherland was the *Turnverein*, an athletic and social (that is, beer-drinking) association that used a gymnasium as its clubhouse. The Missouri Gymnasium was particularly well appointed, with a full set of modern strength and fitness equipment, areas for reading and game playing, and hot and cold baths. Gazing out over the gym floor, the sickly young Bernard swore an oath: "I'm going to be like them. I'm going to look like them."

The price of muscles was steep, however. The initiation fee alone was $15, a sum that would take Bernard months to save. When he started to beg Uncle Harvey to lend him the money, the elder man did what he thought was the only sensible thing: He changed the subject and took his nephew to the zoo.

Fortuitously, someone at the gymnasium had given Bernard a pamphlet of dumbbell exercises. Unable to scrape together the $15, the boy instead went to a sporting goods store and spent fifty cents on a pair of used dumbbells. Armed with his weights and instructions, he retired to his room at Uncle Harvey's house for his first session of strength training. The morning after, he awoke to sore muscles and a new sense of purpose. "Thereafter," he wrote, "I had but one object in view . . . I would not be satisfied until I was a strong man."

After a month, Bernard invested another fifty cents in a second, heavier pair of dumbbells. Each morning, he threw open his windows and hefted iron until exhausted. He worked out for a half hour during his lunch break and walked the six miles to and from work each day. When that proved insufficiently strenuous, he found a ten-pound bar of lead, wrapped it in newspaper and carried it around in his shirt. Each evening upon returning home, he would remove his burden and set it on the hat rack in his family's front hall as another man might leave his briefcase. After witnessing this routine a few times, Grandma Miller was the first, but by no means the last, person to go on record as believing that her grandson might be showing lunatic tendencies.

"You'll be in the insane asylum yet!" she shouted.

Intoxicated by a performance of the Barnum circus (or, to be more precise, by "the magnificent physiques of the male performers"), Bernard and two neighborhood pals constructed their own training gymnasium in Uncle Harvey's cellar, complete with swinging trapezes and a horizontal bar. He taught himself to walk a tightrope and seriously considered a career as a circus acrobat.

The latest health fads also caught his attention, and Bernard attempted to keep up, with mixed results. Always an early adopter of cutting-edge exercise technology, he purchased a high-wheeled bicycle—one of those nineteenth-century cycles with the comically large front wheel—but abandoned it after a series of nasty accidents, the last of which rudely introduced his cranium to a telegraph pole.

A century before Nike's Phil Knight hawked his first pair of running shoes out of his car trunk, America's battle against sloth was being plotted on the playing fields of Harvard. America followed Europe's lead in discovering the power of exercise. Germany and Sweden had added calisthenics to school curricula early in the nineteenth century. The Young Men's Christian Association, or YMCA, was founded in England in 1844 to combat the noxious social conditions that resulted from the mass migration of young men to big cities during the Industrial Revolution. The crowded tenements in which these men lived were Petri dishes for cultivating everything from tuberculosis to venereal disease. The Y and other groups that drove England's "Muscular Christianity" movement sought to save lives and souls by building strong bodies. By midcentury, the notion that a prescription of proper diet combined with regular physical activity could have a deterrent—or even curative—effect on disease had been given a name: physical culture.

Though Americans had a history of importing fitness fads from the continent—Ben Franklin wrote in 1786 that he'd reached a ripe old age because "I live temperately, drink no wine, and use daily exercise of the dumbbell"—it was not until the United States was well into its own industrial revolution that its first homegrown fitness guru emerged. He was Dioclesian Lewis, a Harvard University physical education instructor. In September 1860, *Harper's*

Weekly magazine hailed him as the genius behind the country's "athletic revival." His enemies were stress and inactivity, and his weapon was exercise. The newly formed classes of deskbound office clerks and the expanding ranks of housebound urban mothers were stockpiling nervous energy in their pale, untaxed bodies like pressure building in a Fulton steam engine. In 1869 the physician George Beard gave this malaise the name *neurasthenia* (it was also commonly known as *exhaustion*). Americans were warned to beware its symptoms: insomnia, anxiety, headaches.

The Lewis system of European-influenced calisthenics, as espoused in his 1862 bestseller *The New Gymnastics for Men, Women and Children*, was among the first to outline a program for improving one's life through exercise, a revolutionary idea at a time when complete rest was the preferred cure for stress-related symptoms, and women in particular were often confined to bed and spoon-fed milk. His drills, focused largely on improving flexibility, hand-eye coordination, and physical grace, included beanbag tosses and lifting light weights. Indian clubs, those bowling-pin-shaped training aids now seen only in juggling acts and the dusty corners of school gymnasiums, were popularized by Lewis.

Also working at Harvard in the 1860s was Dr. George Windship, who saw exercise as the route to raw animal strength. This onetime weakling MD, sometimes called "The Roxbury Hercules," was motivated to transform his body through strength training after bullies threw his books down the stairs. Though relatively small in stature, Windship was considered by many to be the world's strongest man, and was said to be able to do pull-ups with only his pinkie grasping the bar. Windship created a popular forty-minute daily strength routine, patented an adjustable dumbbell, and opened a chain of Boston gymnasiums. His system and Lewis's had little in common, but the men did agree on two important points—that America was out of shape, and that exercise was the best route to improving one's health.

On the heels of Lewis and Windship came a third Harvard

man, Dudley Allen Sargent, the "acknowledged Newton" of American physical education, who managed to combine both men's ideals. Sargent, who had a medical degree from Yale, converted Harvard's Hemenway Gymnasium into his laboratory for inventing the Sargent System, which used pulley-weight machines that could be adjusted upward as students grew in strength. Nearly a century later a similar system marketed under the name Nautilus helped kick off the fitness boom in the 1970s.

Sargent was the first to use the term "preventive medicine" in relation to exercise. His work had at least three immediate, related results. It legitimized physical culture as a teaching discipline under a new name, "physical education." It revived the classical ideal of *mens sana in corpore sano* ("a sound mind in a sound body"), which in turn spurred the idea that exercise and sport built character, which in its turn fueled the astronomical growth of school sports, football in particular. Finally, it melded the humanistic ideals of Lewis to the pump-you-up methods of Windship, then married this better-living-through-exercise concept to the late-nineteenth-century vogue for evolutionary thinking. If humankind was still improving, Sargent's reasoning went, exercise could only accelerate one's personal rate of betterment.

Nowhere was this interpretation of "survival of the fittest" made more explicit than in William Blaikie's 1879 bestseller *How to Get Strong and How to Stay So.* Blaikie was a strongman and endurance athlete—he once walked the 225 miles from Boston to New York in a record four and a half days—and a protégé of Sargent's. *How to Get Strong* railed against the sad shape of America's youth and office workers, then offered some exercises (heavily indebted to the Sargent system) to rectify the situation. In an early chapter, he makes a subtle appeal to patriotism when he compares America's boys to the English lads in *Tom Brown's School Days,* who scamper through a nine-mile game of hares and hounds without breaking a sweat. "Let him who thinks that the average American boy . . . would have fared as well go down to the public bath-house, and look carefully at a hundred or two of them as they tumble about in the water. He will see more big heads and slim necks, more poor legs and skinny arms, and lanky, half-built bodies than he would ever imagined the whole neighborhood could produce."

A copy of *How to Get Strong* made its way into Bernard Mc-Fadden's hands in 1883, around the time he wandered into the Missouri Gymnasium. A friend would later compare the physical culturist's initial exposure to Blaikie's theories to the day Thomas Carlyle cracked open one of Plato's works: "He felt as if he were reading his own ideas, set down cogently and clearly, for the first time." Bernard's discovery of Blaikie coincided with the westward progress of the fashion for physical fitness. A story in the August 18, 1881, *St. Louis Post-Dispatch* would certainly have caught the eye of a self-sufficient thirteen-year-old whose hobbies included building trapeze sets and sneaking in extra dumbbell workouts during coffee breaks. The article declared that "the fad for 'physical culture' is developing its fruits, and the girls with well-developed muscles are meeting the men upon their own fields and vanquishing them sometimes, if we may believe all we hear from the summer resorts."

Even before it earned its famous "Show Me" nickname, Missouri was notoriously contrarian—the only state in the Union carried by Stephen Douglas in the 1860 presidential election versus Abraham Lincoln. That tendency to doubt orthodox ideas made Missouri a welcoming environment for new fitness theories. The state was especially receptive to practitioners of alternative medicine. Osteopathy, the science of bone manipulation, was developed in Kirksville, Missouri, by Andrew Still in the years after the Civil War. Chiropractic was invented over the border in southern Iowa in the 1880s, and founder Daniel David Palmer found a welcoming audience among his Missouri neighbors. Two treatments imported from Germany, the hot and cold baths and compresses of hydrotherapy, and the diluted potions of homeopathy, were widespread in the state.

New dietary ideas were also coming out of the health homes and sanitariums east of the Mississippi. Around the time he found Blaikie, Bernard discovered the works of the vegetarian reformer Sylvester Graham, now best known as the originator of the sweet cracker that is an essential ingredient in S'mores, but in the 1800s a powerful advocate for what would a century later be called "health food": fresh vegetables, uncaffeinated beverages, whole grain bread. This last in particular appealed to Bernard. He developed what would become a lifelong hatred for white flour, or "dead food" as he later called it. "I saw that white bread was frequently condemned and I, whenever available, secured whole wheat or Graham bread," he wrote of this rebellious teen-age phase.

After four years in St. Louis, Bernard was dissatisfied with city life. "To be sure, I had secured a valuable business education during this period," he later reflected. "But . . . had I remained on the farm during these four extraordinarily important years in the life of a growing boy I would have undoubtedly added two or three inches to my stature and from forty to sixty pounds to my weight." While his math might have been a little shaky, his conviction was not. Not even twenty or more hours a week in the gym, he felt, could undo the negative effects of city life and forty-plus sedentary hours spent indoors. Only the rigors of farm life would allow him to elude his fate as a consumptive. At age sixteen, after taking time to convince Uncle Harvey that he was doing the right thing (Grandma Miller never came around), he packed up his Indian clubs and lit out for rural McCune, Kansas, where a cousin had founded an outpost of frontier dentistry.

Bernard made the six-hundred-mile journey west alternating between walking and hoboing rides in freight cars. He worked odd jobs along the way, chopping wood and fetching water on a construction site (where "the men were rough and many of them never took a bath"). After several months of this itinerant life-style, he arrived in McCune and was greeted warmly by his cousin E. F. Medearis. Bernard even had a part-time job lined up. He briefly signed on as his cousin's dental assistant, a role which consisted mainly of holding patients in a headlock while Medearis extracted their teeth.

The farm position that Bernard finally secured paid $10 a week, plus meals. The menu was not exactly Graham-friendly— mostly white bread and bacon, which left Bernard to subsist on milk and the occasional serving of navy beans or corn pone. Within three months of agricultural living, however, he felt his strength reviving. His distaste for farm wages also staged a comeback. When the farmer who employed him refused to give him a raise to a full adult salary, $20 a week, he returned to Medearis, who put him to work grinding and polishing false teeth.

Between dental responsibilities, Bernard managed to make two significant additions to his résumé. The first was his initial attempt at teaching athletics, a boxing school he opened in a horse stable. "This school was a success as far as providing entertainment for the idle men about town," he remembered, "but it was a most emphatic failure in adding to my finances." The stable's next occupant, a showman with a two-headed calf, fared better.

The second came about when Medearis—who seems to have shared the entrepreneurial gene with his cousin—hired him to work on a side project, a local newspaper called the *McCune Brick*. (The title referred to his cousin's intent to "hurl advice at the community.") Here he was employed as a printer's devil, or apprentice. Bernard quickly moved up from sweeping floors to running the hand press to sitting in for the hard-drinking compositors during their occasional binges. Unfortunately, the *Brick* sank. Bernard was forced to return to St. Louis.

"I was not far from seventeen years of age when I returned to the city," Macfadden wrote of his arrival in St. Louis after a year away. "My physical difficulties had all disappeared. My ideas were more definitely fixed. I had decided that my future should be spent in the physical culture field."

Bernard's first priority was finding work. Jobs for nonunion compositors were scarce, so he found temporary employment driving a delivery wagon before once again taking a short-term office job. Bernard bounced around some, renting rooms in a series of boardinghouses. A few doors down from Uncle Crume lived a young Greek American named George Baptiste, who worked in his family's awning company.

Baptiste was memorable enough to be the first friend Macfadden mentions by name in any of his memoirs; this may be because when they met, "the mighty fellow was holding out an eighty-pound dumbbell at arm's length on a level with his shoulder, holding it out nonchalantly, as if he had forgotten he *was* holding it out." Baptiste was also, not coincidentally, a member of the Missouri Gymnasium.

Reunited professionally with Uncle Harvey, working as a clerk, and saving every spare cent toward "tuition" at the Missouri Gymnasium, Bernard finally made his first appearance as a member in the requisite garb of that time: a pair of tights and a sleeveless shirt. "When I had arrayed myself in this new costume I felt somewhat strange," he later admitted, "and as I accompanied my friend [presumably Baptiste] from the dressing-room to the gymnasium the emotions that stirred me would be hard to describe."

Though his bookkeeping job took up most of his weekday hours, Bernard's attention was now entirely focused on exercise. He was spending two to three hours a day in the gym, and soon built an impressive physique. A photo of him shirtless at age eighteen reveals a body comparable to that of an Olympic gymnast, with especially well-developed shoulders and upper arms. The comparison is apt, as Macfadden set out to train himself as a horizontal-bar performer and to complete forward and backward somersaults. His legendary confidence in his own strength was developing in tandem with his brawn. "Feats which ordinarily take years to perform," he wrote, "I learned in a few months."

Like many newly converted zealots, Bernard couldn't resist preaching about his cause. One afternoon at Uncle Harvey's, a massive deliveryman who'd grown weary of the sermons tricked Bernard into an impromptu wrestling match by asking him if he thought he could break a particular hold. In front of a crowd of coworkers, Bernard picked up the laborer and heaved him to the floor, managing to tear out the seat of the fellow's pants in the process. For a young man starved of attention, the effect was narcotic.

Bernard and Baptiste were inseparable and worked out together daily at the gymnasium. Years later, living in temporary exile in England, the adult Bernarr Macfadden relived those passionate days. Rounds of calisthenics were chased with long wrestling bouts that left both men "dripping with perspiration." Often a run of one or two miles followed, then hot and warm showers. "When such exercise is not overdone," he wrote, "you seem to be thrilled almost continually with the throbbing, pulsating forces of life."

Because their sport is rigidly organized by weight classes, wrestlers constantly strive to trim as many ounces off their frames as possible without sacrificing strength. Bernard and Baptiste began to experiment with reducing their food intake. Bernard started by exchanging the "slavery" of the traditional breakfast-lunch-dinner meal plan for a two-meals-a-day routine. By his reckoning, this was his first great discovery in physical culture. He would preach its merits for the next seven decades. Astonished by the increase in strength this change provided, he proceeded to pare his daily caloric intake down to one serving of vegetables. After a few weeks on this regimen he had lost twenty-five pounds from an already trim frame. Only after nearly fainting on the street one day did he realize that he'd probably taken things too far. Baptiste felt his friend's pulse and found that it had fallen to thirty-five beats per minute—the body slows its metabolism when it thinks it's starving—and insisted that they end their experiment immediately.

Bernard's professional ambitions continued to grow. At the age of eighteen, he had taken on the added role of bill collector and saved up enough money to enter into a partnership in a laundry, which quickly failed. The brief period of unemployment that followed provided an opportunity for Bernard to combine the business skills that he'd picked up as a bookkeeper and small-business owner with his passion for wrestling. He and Baptiste started organizing tournaments. Bernard found he was naturally gifted at shooting his mouth off to attract attention and lure paying customers.

One of his first successes was coercing a top local athlete into challenging him to a match in the Greco-Roman style, a classical type of wrestling in which one is not allowed to attack below the waist. The combatants were gentlemanly at first, Bernard recalled, but the event soon devolved into chaos when his opponent's mood turned "maliciously hostile." Furious, Bernard picked him up and pinned him against a row of steam pipes along the back wall of the stage. After an hour of such "unsportsmanlike conduct," Bernard's stamina gained him a first fall, and then a second. The theatrical match was much talked about in the following days and had two positive outcomes: The victor became a minor athletic celebrity, and the gate receipts were high enough to inspire the two young promoters to organize more tournaments, drawing champions from both St. Louis and Chicago. Each match boosted both Bernard's reputation as a wrestler and his income.

"There has been no time, probably, when I have felt more satisfied with myself than I did at this period," wrote Bernarr Macfadden in one memoir. About the only thing he wasn't content with was his name, which he found too pedestrian for a man of his abilities and would continue to tinker with over the coming years, in versions such as "Bernie A. McFadden" and "Prof. A. B.

Bernarr." He was earning a decent living as a promoter, had sculpted himself into a perfect physical specimen, and had taken it upon himself to spackle the gaps in his education through hours of study. The Missouri Gymnasium, like most *Turnverein* clubs, had a reading room where Bernard would have been likely to find books and periodicals on diet and exercise. So as not to waste a minute of potential self-improvement, he carried a dictionary everywhere.

In January 1891, Bernard made his first appearance in print, in the *St. Louis Post-Dispatch*, mentioned as a member of the Shamrocks, a local football club. The article noted that an unnamed Shamrock "issued a challenge for [a] game with a pot of $100 to $500." It's not hard to imagine a cocky young wrestling promoter offering such a wager.

Sometime in the spring of 1891, outside his new apartment on what he judged a "fairly prominent St. Louis Street," Bernard hung a shingle that read:

BERNARD MCFADDEN—KINISITHERAPIST
TEACHER OF HIGHER PHYSICAL CULTURE

The term *kinisitherapist* was his own etymological Frankenstein, the sort of big word that might be invented by a young man prone to toting around a dictionary. He intended the term to mean "the use of movements in the cure of disease." Regardless of his intentions, no one in St. Louis had the slightest idea of what he meant. The studio was a modest success, however, in part because Bernard had also printed a sign bearing his snappy new marketing slogan: "Weakness Is a Crime—Don't Be a Criminal." From his work as a trainer and his income from organizing wrestling matches, he was making a comfortable living for a young, single man.

Bernard was an eager dabbler in "irregular," that is to say, not medically sanctioned, methods of increasing one's level of wellness. Not all of his experiments were successful. Hydrotherapy and other "water cures" had made fashionable the notion that a clean colon was essential to good health. Bernard's wide reading in the field of health food introduced him to a variety of edibles that "were valuable when used with other foods that would add bulk or waste elements," which is to say they would empty his bowels with Swiss chronograph regularity. His written recollections of this period are a little vague, but it sounds as though he was attempting to live on a diet of high-fiber laxative. When he began to notice serious problems with his digestive powers, he reduced his intake of the supplement and increased his already large daily ration of exercise. Calorie starved and probably suffering from excruciating gas, he watched his weight decline until he'd lost thirty pounds. Also lost were his vigor and most of his clients, who were spooked by his skeletal appearance. His kinisitherapy career was put on hold.

During his months as a studio operator, Bernard had begun to submit articles to newspapers and other publications, proclaiming the benefits of physical culture. The quality of these writings was not high. "They were returned to me with unfailing regularity," he recalled. Hoping to improve his rhetorical skills, he returned to school, as a student and part-time physical culture teacher at the Bunker Hill Military Academy, about thirty-five miles from St. Louis across the border in Illinois. He traded his labor for room, board, and an accelerated education. The workload was exactly what he wanted—one to two hours of daily instruction, with "the balance of my time devoted to studying my favorite subject—literature."

Bunker Hill was the setting for what Bernarr Macfadden recalled as his "first insight into the possibilities of natural curative methods." For simplicity's sake, Bernard had temporarily abandoned his two-meals-a-day habit for the three squares served at the academy. Near the end of the spring term, he felt the early symptoms of pneumonia coming on, and immediately diagnosed his relative gluttony as the cause. Bernard had noticed that farm animals became well by abstaining from food when ill. He immediately cut his own intake back to a couple of pieces of fruit per day. By the second day of fasting, his chest had begun to clear. By the fourth day, the inflammation had all but vanished. His fitness philosophy was starting to gel: lots of exercise, limited calories, nontraditional cures.

Bernard was invited to return to Bunker Hill, but a year of unsalaried study had left him broke. Instead, he returned to promoting wrestling matches. As before, he exploited both his business skills and his status as a drawing card. He offered prizes of $15 to $25 to the winner ("I always won," he recalled), but his take as business manager averaged $100 to $200. When a tournament of "champions"—a term applied as loosely toward wrestlers in those days as "gourmet" is to foodstuffs today—came to St. Louis, Bernard hatched his grandest scheme yet: He would starve himself down to the lightweight class at 135 pounds, win the tourney, then challenge the winners of the welterweight, middleweight, and heavyweight classes to "private matches which would draw large crowds."

Bernard trained ten hours a day with Baptiste to cut twelve pounds, and when he took the ring against a man called "the lightweight champion of the West," his ribs stuck out like the teeth on a comb, earning him "a nice round of laughter from a packed house." He pinned the champion in seven and a half minutes. The welterweight champion of Chicago and an unknown middleweight also fell.

The promoter now had a problem—the tournament's heavyweight winner had no interest in facing him. So he looked east. Frank Whitmore was the heavyweight champion of Chicago and a captain of that city's fire department. He and Bernard had once wrestled an informal match, which Whitmore had easily won. Since they had wrestled in the catch-as-catch-can style previously (in which no holds are barred), Bernard issued Whitmore a "defi"—the sort of "Here I am, come and get me" challenge familiar to wrestling fans of all eras—in his preferred style, Greco-Roman. Whitmore accepted.

Bernard's potent publicity cocktail—the defi, the rematch, the novelty of his taking on a grappler twenty-five pounds heavier than he—proved irresistible to ticket buyers. Whitmore's pre-match thoughts are unknown, but his young challenger later allowed that as he stepped on the stage before the packed house in one of St. Louis' largest theaters, he felt "as though my whole life depended on this victory."

Considering the aplomb with which he'd disposed of his previous opponents, it is safe to assume that Bernard expected an easy victory against Whitmore, too. This would have been a mistake. Whitmore was still in top form. The two traded the advantage for over an hour. The heavyweight used his size to secure a series of holds that should have allowed him to pin the smaller man to the mat; however, the challenger employed his well-trained neck muscles to keep his shoulders arched above the floor. Finally, after ninety minutes, Bernard pinned Whitmore with a full nelson. A second fall and victory followed a few minutes later. The 135-pound Bernard was now, a biographer later observed, "presumably the heavyweight champion of Chicago, which he had never seen."

Capitalizing on the burst of notoriety from his victory, Bernard barnstormed in and around St. Louis as a popular professional wrestler for a few months. One match in front of more than 2,500 spectators ended with his opponent cheating and audience members storming the ring. Another out-of-town contest went on for over three hours, during which the local fans screamed "Choke him! Kill him!" until the battle ended in a disputed victory for Bernard. He and his companions were forced to run out of town in the dark. The time seemed right for another career change.

His regional celebrity opened a door. In 1892, the now twenty-three-year-old Bernard was offered a position as physical director at Marmaduke Military Academy. This private boys' school had been opened by a consortium of forward-thinking St. Louis businessmen with the goal to "unite careful scholastic training with physical vigor and moral tone." Bernard didn't hesitate to sign on. His salary was $50 a month, plus board. "Professor B.A. McFadden, the champion wrestler," as he was identified in a local newspaper, taught grappling and coached the football team with great success. The rules governing school athletics were apparently as fuzzy as those sanctioning wrestling champions. Coach McFadden not only quarterbacked the football squad but boasted of importing a "big burly half-back" from Illinois for a game against a hated rival. His students, who topped out at fourteen years of age, played the University of Missouri B team to a 6–6 tie.

Bernard's year at Marmaduke left him ample opportunity to finish his first serious literary effort, an 80,000-word novel entitled *The Athlete's Conquest*. The plot, if it can be called that, revolved around a young man's desire to find a physical-culture bride. Decades later, the author would advise one of his daughters that to be a great writer one should read Jane Austen. The opening lines of *The Athlete's Conquest* betray more than a little influence of the first sentence of *Pride and Prejudice*:

Austen: "It is a truth universally acknowledged, that a single man in possession of a good fortune must be in want of a wife."

Macfadden: "Here I am, twenty-seven years of age, with a good income, a fine business and everything necessary to make one happy—except a wife."

His attempts at copying Austen's style go downhill from there, not least because his novel's message—physical culture is good; wearing corsets is bad—takes precedence over story, character development, and pretty much anything else that makes fiction worth reading. "I had not as yet formed the really necessary habit of pruning and correcting," the author conceded. "Consequently it can be imagined that the manuscript was unusually crude and unpolished, and that many grammatical mistakes appeared therein." The above excerpt is taken from the 1899 reissue, which had been worked over by a team of professional editors. One wonders how painful reading the original would have been.

As was usually the case with Bernard, when competence butted heads with confidence, the latter carried the day. He had his manuscript typed and dropped off at a publisher. The author stopped by a couple days later to ask how the publisher's reader had liked it.

"He does not consider it fit for publication," said the manager.

"What is the matter with it?" I asked.

"To begin with, it has no plot, and then it is poorly expressed, crude and ungrammatical."

Bernard revised *The Athlete's Conquest*, then paid to have it published. It is a testament to his popularity at Marmaduke that every student purchased a copy. When one boy's book was confiscated because he was discovered reading it after lights out, the author proudly bestowed another on him.

Bernard considered his year at Marmaduke a great success in all ways but the usual one, financially. When classes ended, he spent a few weeks giving wrestling and boxing lessons in a nearby town. Here he took stock of his life. Bernard McFadden was now twenty-four, adequately educated, supremely self-assured, and incredibly ambitious. For a young Midwesterner in possession of those qualities in the summer of 1893, there was only one place to be: The Chicago World's Fair.

Chapter 3

Go East, Young Man

I was always gifted with an unlimited amount of self-confidence. You get a rude awakening now and then when you are oversupplied with this characteristic, but it is better to have too much than too little.

—BERNARR MACFADDEN, "MY FIFTY YEARS OF PHYSICAL CULTURE," *PHYSICAL CULTURE*, SEPTEMBER 1933

To a visitor arriving from a year in rural Missouri, the Columbian Exposition must have been staggering. Bernard McFadden came to Chicago by boat. His first steps in the city took him onto the fair's moving electric sidewalk. This marvel conveyed him to the Peristyle, a massive Beaux Arts structure a hundred feet high, ornamented by gigantic nude statuary. Beyond that lay the fourteen spectacular buildings of the White City, master planner Daniel Burnham's homage to the world's great monuments of architecture. Every square inch of the 633-acre fairgrounds seemed to be packed with treasures and curiosities: the wonders of the Electricity Building, home to novelties such as moving pictures and the first electric chair, brilliantly lit at night with thousands of Thomas Edison's newfangled light bulbs; the actual Liberty Bell; an eleven-ton wheel of Canadian cheese; a giant medieval knight on horseback sculpted entirely from prunes.

Just outside the main fairgrounds, across Stoney Island Avenue, lay the entertainments of the Midway. Here could be found the world's first Ferris wheel, a 250-foot-high contraption forged out of two hundred tons of iron that spun 2,160 riders through the sky. Mixed in among the oom-pah-pah bands of the German Pavilion and the exotic "hootchy-kootchy" dancers of Egypt (many of whom looked suspiciously Midwestern) were hawkers pushing the world's first hamburgers, carbonated beverages, and Cream of Wheat.

Bernard gravitated almost immediately to an attraction at a nearby theater, a vaudeville act put on by a pair of entrepreneurs just a year older than himself, each of whom was at the onset of a brilliant career as a showman—Florenz Ziegfeld and Eugen Sandow.

Florenz Ziegfeld would make his fortune as the impresario behind the over-the-top Ziegfeld Follies, the jazz-era musical comedy revue. But in 1893, he was the anonymous son of the owner of the Trocadero, a struggling Chicago nightclub. When Ziegfeld the elder sent his son out to find an act that would rescue them from bankruptcy, he returned with a strongman. Sandow, a Prussian born in 1867 as Friedrich Wilhelm Mueller, was already one of the world's most famous athletes. No less an expert than Harvard's Dudley Allen Sargent had proclaimed him "the most wonderful specimen of man I have ever seen." He was also a gifted showman, as a rave review from the *Chicago Tribune* made clear:

Sandow . . . picked [a man] up by the most convenient part of his garments, and then, after juggling the poor fellow in the air, he threw him away just as a little boy does the kitten when it scratches him. He next took a fifty-six pound dumbbell in each hand, and, blindfolded and with his feet tied together, actually turned a back [flip]. . . . After some great feats . . . Sandow form[ed] an arch with his body, breast upward, resting himself on his hands and feet, a huge teeter-board was balanced on his chest and then three horses were led onto the board . . . where the animals teetered up and down, finally standing still directly over the wonderful chest of the man. And when Sandow stood upright after this great feat he didn't show half as much weariness as an ordinary man does after fetching his wife a pitcher of water.

Sandow cannily marketed himself to late-Victorian audiences on two continents through photographs of his muscular physique ornamented only by a fig leaf. At the apex of his fame in America in the 1890s, he was earning $1,500–$2,500 a week for his strength demonstrations.

In keeping with the neoclassical theme of the Columbian Exposition, Ziegfeld had Sandow emulate famous statues. Dusted with white powder, the strongman struck poses of Hercules, Samson, and Ajax while positioned in front of a black velvet cabinet. Audiences were mesmerized by the sight of Sandow's massive fifty-two-inch chest contrasted against the dark background. No one was more impressed than Bernard McFadden, who filed the trick away in his memory. It would not be the last idea he would borrow from Sandow.

At the fair, Bernard met up with a friend. Alexander Whitely was the inventor of the popular Whitely Exerciser, a wall-mounted resistance trainer built of ropes and pulleys. The exerciser was marketed to the growing numbers of business travelers and other potential victims of overwork, stress, and "nervous energy." Bernard's endorsement as physical director of Marmaduke Academy had appeared in Whitely advertisements. Bernard bankrolled his summer in Chicago by demonstrating the Whitely Exerciser on the Midway, and soaked up as much of the fair as he could in his off hours. When October arrived, he returned to St. Louis with a hundred dollars in his pocket and a head full of new ideas.

Bernard had realized that St. Louis was not big enough for a man with his ambitions. "I was convinced that I should settle down in some city where my work would be appreciated," he later wrote. "I did not feel like St. Louis would offer me the opportunities to which my work justly entitled me, and I at length made up my mind to go to Boston." The choice made perfect sense for an aspiring writer and kinisitherapist. Boston was the undisputed center of American literature and culture, not to mention the home of Harvard, the cradle of physical culture civilization. The sage of Boston, Ralph Waldo Emerson, might as well have been whispering in Bernard McFadden's ear these words from his essay "Self Reliance" as the muscular young man boarded his train in St. Louis carrying a suitcase filled with dumbbells: "To believe your own thought, to believe that what is true for you in your private heart is true for all men—that is genius."

Bernard's rail route to Boston took him via New York City, where, in those pretunnel days, passengers detrained on the west side of the Hudson River and caught a ferry to Manhattan, from which they continued northward. His train arrived at the Jersey City station in the evening. He caught his first glimpse of Manhattan as the city's ten-story skyline, culminating in the fourteen-story *New York World* building, twinkled seductively across the water. He decided to linger a few days before continuing.

Like countless new arrivals in Manhattan before and after him, Bernard determined that if he could make it there, he could make it anywhere. He may, however, have been the first immigrant to count among the island city's finest attributes the relative corpulence of its populace, which seemed a promising customer base. He was also stunned by the cleanliness of New York City, which had banned the burning of the soft coal that left St. Louis mired in soot. He had $50 to cushion him. Boston could wait.

Bernard rented a two-room apartment on the first floor of a rooming house at 24 East Twentieth Street, just off Broadway, for $10 a week. He printed photographs of himself shirtless "in classical poses" on a thousand circulars and posters promising a free "Physical Culture Matinee" the following week. It is likely that those advertisements were the first to feature the new, catchier spelling of his name: Professor Bernarr Macfadden. Bernard was an ordinary name, he explained. "I was starting out fresh. New business, new kind of culture, new name—the works." He also settled into the public image that would mark him for the rest of his life—wild-haired, sloppily dressed, shoeless and hatless if possible. He was a firm believer in the noble savage and took pride in looking like a Native-American brave, scanning the horizon for something to scalp. He would, of course, show off his arms, thickly sinewed with muscle, to anyone who asked to see them.

Among those attracted by the flyers was a reporter from the *New York Sun*, who wrote that Macfadden "chatted and posed in an interesting way for more than an hour." Neither the publicity nor the demonstration brought in any paying customers, though. As Macfadden neared the end of a fortnight, his funds were almost exhausted. "You could not secure favors from New York landladies," he wrote. "I fully understood that I had to pay my rent or get out."

The day was saved when a prosperous actor walked in and asked for a price quote to get whipped into shape. Macfadden opened with a bid of $50, expecting to haggle, but the actor pulled out five $10 bills. Other students followed, skinny men who wanted to bulk up and fat ones who wanted to pass their life-insurance exams. Soon Macfadden was living comfortably, supplementing his income with odd jobs: selling exercisers for Whitely, teaching wrestling at the Manhattan Athletic Club, and modeling in the nude for the Art Students League.

Macfadden's appetite for publicity, and his skill at getting it, seem only to have magnified with his move east. He arranged to get his promo shots into the hands of Napoleon Sarony, New York's top celebrity photographer, who agreed to shoot him. Today Sarony is best known for his pouty portraits of Oscar Wilde, but Macfadden probably knew a popular series he'd done of Sandow in 1893–94. Macfadden fasted for a week prior to the photo session to accentuate the definition of his muscles. Rather than feeling debilitated after consuming nothing more than water for days, Macfadden was able to hold a one-hundred-pound dumbbell aloft for the fifteen seconds needed to make a photographic negative. As an encore, he allowed Sarony to capture him lying on his back, chest-pressing a two-hundred-pound man.

Macfadden was bored with the life of a personal trainer, even after moving uptown to a more "swanky" (his description) location at 296 Fifth Avenue. He continued to scribble articles on the topics of "body-building and curing by nature's methods," but all save one or two were rejected. His work with Whitely had made him aware of the expanding market for home exercisers; he developed his own, "a combination of rubber and cords running over pulleys," and contracted with a sporting-goods manufacturer to sell them. The company went bankrupt within months. With his typical resilience, Macfadden simply made a few design improvements and fabricated the exercisers himself. And when sales in the United States proved disappointing, he packed some samples and boarded a Cunard liner bound for England.

"A change of environment is an extraordinary stimulus to original thought," wrote Bernarr Macfadden of his journey to England in early 1897. A rocky voyage in second class made him so seasick that he vowed never to return to America once he stepped off the ship in Liverpool. He may have altered his name in America, but it was in England that he truly became Bernarr Macfadden.

As soon as his stomach had settled, Macfadden made arrangements to have his exerciser manufactured and distributed. Almost simultaneously, he embarked on a lecture tour to market the apparatus. His timing was excellent: The late-Victorian appetite for self-improvement had spurred a mania for physical fitness that had spread from London throughout the kingdom.

Once again, Macfadden found himself following Eugen Sandow's example. When the Columbian Exposition ended, the Prussian had moved to England and cashed in on his popularity by launching a chain of physical-culture studios, publishing strength-training books, and giving more of his trademark strongman demonstrations.

Though Sandow would have dwarfed Macfadden in a side-by-side comparison—at five six and 145 pounds, Macfadden was built like a three-quarter scale model of the Prussian—Macfadden's energy and persistence gave him a huge advantage as a salesman. He moved from town to town, pasting up posters and taking out small newspaper advertisements. Each performance commenced with Macfadden, dressed in only a loincloth, executing the same artistic poses he'd seen Sandow perform in Chicago. Positioned in front of a cabinet lined with black velvet and lit from below, Macfadden's rippling physique appeared twice as large as it actually was. Once he'd dazzled the crowd with his posing, he followed with a lecture on physical culture and a demonstration of his exerciser.

Macfadden had been nervous about the reception his Ozark twang would receive in Britain. His concern was unnecessary: The show and the apparatus were instant hits. "There was a far more keen appreciation of my message in England than there was in my own country," he wrote. He began calling on volunteers from the audience to come up and prove their manliness by using the exerciser in one-on-one strength contests. He hired a full-time assistant and secretary to keep up with the demand for lectures.

Toward the end of this first tour, Macfadden had another brilliant sales idea. Perhaps inspired by the life-changing pamphlet of dumbbell exercises he'd received on his first visit to the Missouri Gymnasium, he published a four-page brochure describing how to use the Macfadden Exerciser. When British audiences devoured this fitness wisdom, the frustrated writer began adding his rejected health articles to the publication's contents. What started as an instruction manual was soon transformed into a miniature physical-culture periodical; inside was an address to which interested parties could send money to receive future editions. Orders flooded in. Macfadden quickly arranged to publish the pamphlets regularly out of London as *Macfadden's Magazine.*

Macfadden was always imprecise with the details of this career triumph. The obfuscation may have been deliberate. Around this time, he married a woman named Bertha Fontaine, whom he'd met while the two were demonstrating exercise equipment in a Brooklyn department store. Among the few things known about Fontaine are that: a) her marriage to Macfadden was quickly annulled; b) Macfadden never publicly admitted that they'd been married; and c) she resurfaced decades later in Los Angeles around the time that Macfadden's political career was unfolding, and threatened to tell the press about their marriage unless she received hush money monthly. Macfadden paid up.

What *is* known for certain is that in July 1898 Eugen Sandow began publishing a fitness magazine in England named *Physical Culture.* Coincidentally or not, by the end of 1898, Bernarr Macfadden had returned to New York with his own idea for a new publication, a health magazine for Americans. He would call it *Physical Culture.*

Chapter 4

Strong Opinions

Every disease in the human body is simply an endeavor on the part of the body to correct an abnormal condition. . . . It is the presence of impurities in the blood that make the production of a cold possible. . . . Disease germs consume these poisons, or render them harmless.

—BERNARR MACFADDEN, *PHYSICAL CULTURE*, MAY 1901

Consumer magazines have traditionally sprung from two sources. There are those started by publishing companies in order to make money. And there are those started by relative publishing novices who want to shout out their message to like-minded souls. Publications in this second group are generally the fruits of a single, monomaniacal vision, which is why they are often the freshest, most timely, and ultimately most influential publications of their day. The *New Yorker* started as such a magazine, as did *Playboy*, *Rolling Stone*, and *Martha Stewart Living*. So did *Physical Culture*.

Volume 1, number 1 of *Physical Culture* appeared in March 1899, bearing cover images of a muscular young man identified as "Prof. B. Macfadden in classical poses." Its price was a nickel. The editorial offices were listed on the table of contents as 88–90 Gold Street, New York City. Nowhere was it mentioned that those world headquarters consisted of a desk in a real-estate office near the Manhattan side of the Brooklyn Bridge.

The new magazine aspired at first to be "little, if anything, more than a catalogue" for Macfadden's exerciser. (The apparatus now came in four styles, priced from $2 to $5.) The issue weighed in at a slim twenty-five pages. The two main articles were titled "The Development of Energy, Vitality and Health" and "Can a Weak Mind Be Made Strong?" *The Athlete's Conquest* also reappeared, as the first chapter in a serial that occupied the second half of the magazine. The entire contents of the debut *Physical Culture* were penned by its editor under various noms de plume, both masculine and feminine. The total cost for producing the issue was $200.

Having pushed his first hatchling out of the nest with the help of creative recycling, Macfadden began to find his voice with his second issue. A summary of his health philosophy appears on page one.

It is the editor's firm and conscientious belief that weakness is a crime.

That one has no more excuse for being weak than he can have for going hungry when food is at hand.

That disease is not sent by divine providence, but is a result of the victim's own ignorance or carelessness.

That [the editor's] great purpose in life is to preach the gospel of health, strength, and the means of acquiring it.

That the finest and most satisfying results that can be acquired from proper physical culture are the cure of disease and the development of that energy, vitality, and health essential to the success and well being of life.

It was a message almost a hundred years ahead of its time: If you are sick or feel weak, don't just sit there and wait for someone to cure you—get off your butt and do something about it. For Macfadden, *Physical Culture* was the pulpit for an evangelical movement—to save bodies, not souls. "To my mind," he wrote in an early issue, the salvation of America's corpus was a mission "of far more importance than any religion. *It is a religion.*" He called on his "converts" to enlist as missionaries:

Who will join our army as *disciples of health*, and fight with us for the annihilation of the horrible curses of humanity?
1. Prudishness
2. Corsets
3. Muscular inactivity
4. Gluttony
5. Drugs
6. Alcohol

This was his Martin Luther moment, and this list of "curses"— soon to be joined by a seventh, tobacco—was his 95 Theses, hammered to the doors of anything that stood against improving health via natural methods.

The first few issues of *Physical Culture* were crudely written and simply designed—by modern standards, they have more in common with a PTA newsletter than a glossy fitness magazine— but more than a century later, Macfadden's personality still leaps off the pages (largely thanks to his generous use of CAPITAL LETTERS, **bold type,** and exclamation points!!!!). The magazine and its editor were both quite openly works in progress, and readers were invited to join in the fun through Macfadden's personal advice column, in which he offered to answer all reader inquiries by mail.

Between 1830 and 1890, eighty-five health magazines had been started in the United States. Almost all of them offered dry, impersonal instruction in the proper denial of earthly pleasure. Macfadden's message was a mix of sermon and confession. Every aspect of his daily routine was shared in detail: his need for eight hours of sleep, his two-meal schedule, his ever-changing exercise regimen, his hatred of vaccination and corsets, and his personal failings and attempts to overcome them. Dietary and do-it-yourself medical advice was offered on nearly every page of *Physical Culture*. Exercises were described in steps, like recipes in women's magazines, and often illustrated by photos of a shirtless Macfadden. If the beefcake shots of the editor didn't inspire readers, Macfadden's tally of his own strength surely did: At age thirty-two he could hoist a hundred-pound dumbbell overhead eleven times, tear a deck of cards in half twice, and do twenty one-legged deep knee bends.

Almost from the start, Macfadden eclipsed his competition, including Sandow's *Physical Culture*, by adding three elements to his magazine that gave it mass appeal. He freely used celebrities, he sought women readers as well as men, and he seasoned every issue with a healthy dose of sex. Sandow himself appeared in the August 1899 issue, writing "Familiar Feats of Strength and How to Do them." In the seventh issue, Macfadden introduced a special section for women, overseen by Dr. Ella A. Jennings, the former editor of the wellness journal *Humanity and Health*. "There can be no beauty without fine muscles," Macfadden explained.

Macfadden put out those first few issues largely on his own, working until eleven or midnight six days a week and most Sundays, too. His amateurish but passionate product found an audience. Sales grew respectably from 5,000 for the premiere issue in March 1899 to 10,000 in October, 16,000 in December, and 25,000 in January 1900.

Those initial volumes drew enough interest to boost exerciser sales, which in turn paid the printing bills. Macfadden was a businessman who operated from his gut, and after a year, he sensed the greater potential of the magazine if he dropped the exerciser promotion in favor of improved editorial content. He referred to this revamped product as "the first issue of PHYSICAL CULTURE Magazine as a real magazine." Sales surged again with the improvements. The early years of *Physical Culture* would see the bylines of writers as diverse as Charles Eliot, the famously outspoken president of Harvard and one of America's great public intellectuals; bare-knuckle heavyweight boxing champion John L. Sullivan; hatchet-wielding temperance advocate Carrie Nation (writing about another evil, tobacco); as well as various promoters of vegetarianism, fruitarianism, naturopathy, fasting, antivaccination laws, bodybuilding, and, in the case of one cape-wearing Spanish count, grass eating. The nobleman claimed his habit prevented nervous breakdowns and stomach troubles, though only if the blades were "well masticated."

About the only health subjects not explored in *Physical Culture* were anything practiced by medical doctors, or allopaths, as Macfadden preferred to call them. When President William McKinley was assassinated in 1901, Macfadden wrote a scathing editorial attacking the fiends he saw as the true perpetrators of McKinley's early demise: his physicians, who fed the president a slice of bread dipped in beef juice too soon after the shooting. "May god forgive the fools!" he wrote. "Another martyr to the cause of medical experimentation."

As *Physical Culture* grew, so did Macfadden's staff. At the end of three years, fifty employees were working out of an eight-room office. Perhaps the most important of these hires was John W. Coryell, a versatile and prolific writer who published popular crime novels under his own name and women's romance novels as Bertha M. Clay. Coryell's appeal to Macfadden is not as odd as it might sound. The character Nick Carter, who narrated Coryell's best-known tales, was a brilliant crime solver, a five-four powerhouse whose father "had made the physical development of his son one of the studies of his life" and who could "lift a horse with ease." Coryell brought a punchy, pulpy style to subject matter not famed for its readability. His first article, "Nudity and Purity"—which posited that the former was the natural state in which the latter would flourish—appeared in the June 1900 issue.

After less than two years as a publisher, Macfadden had a hit. He was not restrained in boasting about it. "The success of this magazine has been more instantaneous and its growth more phenomenal than any other publication ever placed upon the market in this country," he wrote. His willingness to acknowledge nakedness and sex, and to even offer—primarily through books he authored and sold through arresting full-page ads in *Physical Culture*—advice on how both parties might actually enjoy the procreative act, surely didn't hurt sales. (*What a Young Husband Ought to Know*, for example, promised to divulge "three theories of coition.") When a YMCA director in Easton, Pennsylvania, complained about the inclusion of steamy articles in *Physical Culture*, Macfadden lashed out against his prudery. "One of the principal causes of weakness and ugliness at the present time is a lack of respect for the human body—is the idea that it is something vulgar to be hid and despised. Why should we be ashamed of our bodies?" A senior vice president of the Philadelphia and Reading Railroad Co.—one of the world's largest companies and a monopoly so powerful that Theodore Roosevelt's trust busters were driven to dismantle it—wrote to complain about the immoral displays of flesh in *Physical Culture*, and to say that he was removing it from his railway's newsstands. Macfadden responded with a taunting editorial that implied such a pathetic weakling didn't deserve to sell his magazine.

His self-confidence was well-founded. The same issue announced that circulation had grown to 110,000 copies per month. Combined profits in 1902 and 1903 were $90,000, about a million dollars in today's money. Sandow, meanwhile, had changed the name of his magazine to *Sandow's Magazine of Physical Culture*, and would soon truncate it to *Sandow's Magazine* before going out of business. In two years, Bernarr Macfadden had catapulted himself from a loudmouthed nobody to one of America's leading health experts.

By the time New York rang in the twentieth century with a fireworks display over City Hall, *Physical Culture* had migrated north to new offices in the Townsend Building at Twenty-fifth and Broadway. Macfadden had several medicine balls in the air. He'd started a second magazine, *Women's Physical Development*. He was writing advice books after hours; in 1900 and 1901 he was churning out a new title every few months: *The Virile Powers of Superb Manhood*, *Strength from Eating*, and *Macfadden's New Hair Culture*, which described how he defeated the premature onset of baldness at age twenty-six by tugging violently at his hair for several minutes a day. He even managed to squeeze in a second novel, *A Strenuous Lover*, a tale every bit as tiresome as the title suggests. At one point Macfadden was working so hard that he nearly lost his sight from eyestrain. Instead of visiting an ophthalmologist, he made his own diagnosis: too much time at his desk, not enough physical activity. He devised a regimen of ocular exercises to cure himself, then swiftly published *Strong Eyes: How Weak Eyes May Be Strengthened and Spectacles Discarded.*

Macfadden may have been cocky, but he also longed to have his crusades against prudery, vaccination, and white bread taken seriously by the mainstream press, to whom "our theories on the cure of disease were considered amusing," he groused.

Macfadden issued a defi to the entire medical establishment: he would cure, without charge and by natural methods, sufferers of chronic ailments whose physicians had given up on them. The Macfadden Health Home opened in Highland, New York, about fifty miles north of Manhattan on the Hudson River, in early 1901. "Among the thousands of applicants that applied for admission," its proprietor boasted, "we finally accepted one patient each from the following diseases: Asthma, Consumption, Rheumatism,

---

*Yes, it sounds fishy, but a woman named Peggy Hall wrote to me: "When I was in high school in Oakland, Calif. my view of the clock was blurred. My mother sent for the little book on eye exercises and made me do them every night. At 78 I still do not wear glasses, except reading ones for the phone book."

Vital Depletion [apparently some sort of malnutrition], Paralysis [meaning cerebral palsy], and Emaciation." Photos of these six sufferers, along with their sorrowful tales of inadequate treatment from conventional healers, were published in the May 1901 issue of *Physical Culture*.

A few months later, Macfadden was able to report that four patients had responded strongly, especially an asthmatic postmaster from rural Ohio who had arrived "little more than a living skeleton" who "literally fought for every breath he drew." Following one to two days of fasting and a month of exercise and "wholesome food," this wretch had been cured of his breathing troubles and gained twenty-eight pounds in the process. When the article about his miracle cure appeared, so many letters arrived at the postman's home inquiring about his treatment that he demanded that *Physical Culture* pay for a stenographer to answer them. Macfadden called him an ingrate but provided the secretary anyway.

In 1902 Macfadden took a one-year lease on a sixty-five-room hotel on the shore of Long Island's Lake Ronkonkoma and opened a second health home, this time for profit. On the heels of the prior year's well-publicized successes, more patients arrived than Macfadden could house. A tent village was erected to handle the overflow. Those involuntary campers were not pleased to learn that they were paying the same rates as guests with a roof over their heads.

The waterfront location was ideal for swimming and boating, but also for the breeding of mosquitoes, which infested the tent village night and day. ("Never in all my experience have I seen such a multitudinous number of hungry insects," Macfadden later recalled in wonder.) Macfadden was mostly an absentee proprietor, which was unfortunate, as his staff was largely populated by alcoholics who neglected to show up for work, and by thieves who did so diligently. They emptied the safe that held guests' valuables, and Macfadden had to reimburse the losses. After less than seven months, Macfadden threw in the towel.

"When I closed this institution it was with the definite conclusion never to try it again," he wrote.

What exactly was Macfadden up against in his health crusade? The American diet of one hundred years ago was, mirabile dictu, even worse than our modern one. In his history of American eating habits, *Revolution at the Table*, Harvey Levenstein describes how the seeds of America's obesity problems were planted between 1870 and 1914. It was during this span that methods of mass production and shipping transformed the American food industry. Until the late 1800s, the nation's diet consisted largely of salt pork, bread, and potatoes—and plenty of them. Any green matter that snuck into the digestive systems of the average urbanite was probably mold, one reason why constipation was the obsession of nineteenth-century health reformers. The percentage of the American workforce employed as sedentary clerks multiplied fourfold within two generations. As manufacturing improved, prices of processed staples, such as white flour and sugar, fell. Their consumption skyrocketed. It's no coincidence that America's first great weight-loss expert, Horace Fletcher, emerged around this time. The essence of Fletcherizing, as his system was known, was simple: Cut back on protein, meat especially, and chew your food to a paste.

Macfadden, who equated white bread and candy with poison, set out to demonstrate that healthy foods were within anyone's budget, embarking on a one-month experiment to prove that a man could live happily on five cents a day. The regimen had mixed results, since even he could adhere to a diet that consisted mostly of oats, beans, cornmeal, and turnips for only eighteen days. With the need for cheap, healthy meals fresh in his mind, however, he took aim at his next target, the food-service industry. The Physical Culture Restaurant, opened that same year at the intersection of Pearl Street and City Hall Place (a location "as near as we could [find] to the Bowery," the Main Street of New

York's down-and-out community), was his attempt to show that wholesome food could be served inexpensively. A full menu of healthy foods—vegetarian pea soup, whole wheat bread with butter, steamed oats—was available on the first floor for a nickel. A more spartan penny menu was on offer in the basement.

What began as a laboratory for healthy living—the words PHYSICAL CULTURE EXPERIMENTAL RESTAURANT were painted on the front window where one might expect to see something like MOM'S PLACE: HOME COOKING—netted $200 in the first month. Within two years, Macfadden was advertising stock in the Physical Culture Restaurant Company (often alongside flattering stories about the eateries), which by then operated fourteen restaurants in New York, Boston, Philadelphia, and Chicago.

In the Spring 1903 issues of *Physical Culture*, Bernarr Macfadden offered a challenge to his readers. "In nearly every county in these great United States there is held an annual fair in which prizes are offered for the best specimens of the various domestic animals—horses, cows, and pigs," he noted. "But never on a single occasion has a prize been offered for the best specimen of man or woman." He would present a $1,000 award to "the most perfect specimen of physical manhood" that could be found.

Macfadden was assembling America's first bodybuilding show, to be held at year's end, during the inaugural Physical Culture Exhibition. This was another idea pinched from Sandow, who'd held a wildly popular "physique competition" at London's Royal Albert Hall in 1901. Fifteen thousand attendees had admired Britannia's most impressive muscles. Sir Arthur Conan Doyle, the creator of Sherlock Holmes, was among the celebrity judges.* Within months, Macfadden held a "Monster Physical

---

*When I asked the sports historian Terry Todd why every part of these bodybuilders' physiques seemed the same as modern strongmen except for their narrow chests, he told me, "Because the bench press hadn't been invented yet."

Culture Meeting" at the New York Opera House on June 1, 1902. When three thousand fans came out for the event, many were turned away because of lack of seats.

As 1903 progressed, the new contest grew more spectacular in every issue of *Physical Culture*. In May, the competition was expanded to include women, who would battle for their own $1,000 prize. By June the event had escalated into an international invitational. Separate competitions would be held in New York and London, with the finalists from each country meeting in an international physical culture smackdown at Madison Square Garden.

The extravaganza took place between December 28, 1903, and January 2, 1904. Contests included "races for men and women, fencing championships, wrestling bouts and several fasting competitions." The exhibition's grand finale was the judging of the perfect physical specimens. Macfadden saw this as the athletic climax to his personal Olympiad; a Minneapolis newspaper was probably closer to public opinion when it dubbed this unveiling "the shape show." Emma Newkirk was crowned the women's champion. The men's was Al Treloar. A short film of the two winners survives, shot by Thomas Edison's movie production company. In it, Newkirk—who for unexplained reasons (modesty perhaps) is identified as "Miss Marshall"—poses to show off her grace and balance. Judging from her dainty posturing, the "Nutcracker Suite" might have been playing as the soundtrack to her routine. Her skintight white bodysuit, accessorized with a sash tied tightly around the waist, leaves little to the imagination, and with the exception of her most private parts, she might as well have been rolled naked in flour from the neck down. Her legs are stout and strong, like a sprinter's.

Treloar appears in leopard-print briefs and gladiator sandals. He poses in a style more familiar to bodybuilding competitions today, flaunting his creviced abdominal muscles, thick triceps, and broad back. (He'd likely picked up a few pointers at the Columbian Exposition in Chicago, where he apprenticed with

Sandow.) After the exhibition, Treloar moved to California to become Director of Physical Education at the Los Angeles Athletic Club in 1906, a position he would hold for forty-two years. At the LAAC he would train weightlifting champions and play a leading role in introducing the body beautiful to Southern California.

The Physical Culture Exhibition lost several thousand dollars, but the publicity Macfadden received was priceless. Reporters found him irresistible. A Sunday *New York World* feature on the growing popularity of vegetarianism that bore the headline "If There Was Meat Famine—What Then?" prominently featured Macfadden as a spokesman for the new healthy lifestyle, which he often prescribed for the obese, neurasthenics, and those whose passion for alcohol was inflamed by indulging their carnivorous urges. "When I started my physical culture experimental stations in New York I was assailed on every side," he told the paper. "However, such is the history of every crusade, and I have lived to see this phase of the anti-meat movement succeed beyond my sanguine hopes."

Riding the momentum of his new notoriety, he hit the road for a series of lectures in 1904, traveling through Cincinnati, Dayton, Indianapolis, Chicago, Minneapolis, St. Louis, Hartford, New Haven, Providence, Boston, Toronto, and Montreal. At each stop, he gave two free speeches: "The Cause and Cure of Weakness" and "Is Medicine the Science of Guessing?" He charged admission to hear the racier topics, "The Cultivation of Perfect Womanhood" (delivered to female-only audiences) and "The Complete Powers of Superb Manhood" (for men only). His readers heeded his call to form "a physical culture society in every city in America," and by 1905 there were seventeen such health clubs. Back in New York, he opened a public coed gymnasium on February 1, 1905, the Bernarr Macfadden Institute of Physical Culture, in the same building that held his new offices, 29 East Nineteenth Street. "Special evening classes" were advertised, in which men and women could perspire together in the pursuit of fitness.

In the spring of 1905, Macfadden announced plans for his second Physical Culture Exhibition. The sequel was more ambitious. Male contestants could compete in seventeen different decathlon-style running and jumping events, plus some Macfadden originals, such as "throwing a 56-pound weight for height" and sprinting fifty yards while carrying one hundred pounds on one's back. Women competed in eight events. Gold medals were presented to both sexes in each competition, and points tallied. When these totals were computed, factoring in a hefty bonus for "symmetry and general beauty of form," one man and one woman would be named winner of the Grand Competition.

By the end of September, New York City's barbershops and drugstores were plastered in posters for the upcoming show. The advertisements featured striking photographs from the last exhibition. Al Treloar's manly assets taunted patrons and passersby; Emma Newkirk and a dozen or so other perfectly formed women made it clear that admirers of the athletic female physique would not be disappointed. To guarantee that the audience could view the contestants' physical glory from every possible angle, Macfadden had a revolving pedestal constructed that could hold twenty to twenty-five athletes at one time. It was not the only special step he took to ensure that the view of the female competitors' assets was unobstructed. "At that time, of course, to pose bare legged was inconceivable," Macfadden wrote of the Edwardian mores in women's fashion that made a flash of a woman's wrist a thrilling display of skin. "But I found a very economical way of furnishing costumes for the young women in a certain brand of underwear that was quite popular at that time."

Preshow publicity was heavy. On October 2, 1905, the *New York Times* ran a story on the preliminary elimination heats being held on Nineteenth Street, headlined SO CALLED BEAUTY CONTEST: "Physical Culture Show" Begins in Madison Square Garden. The reporter seems to be straining to stifle giggles in his reports on the strength circus underway, a quantity of muscle on display so great that "it bulged out as if it would rend asunder the enfolding skin." The *Times* man noted that the preliminaries had drawn hundreds of would-be competitors, including "a Swedish woman, a well-known artists' model; a black-eyed sylph with muscles of steel, a fighting eye, and a brogue; [and] a chunk of a woman with muscles like those of a man."

Also taking keen notice of the upcoming show was Anthony Comstock, a portly party pooper who headed a quasi-official civic group known as the Society for the Suppression of Vice. While Macfadden was building his reputation as America's leading proponent for displaying skin in public, Comstock had already become a household name as America's most smugly upright citizen. He'd made a career of haranguing flesh peddlers and sexual libertines, and his prominent belly and muttonchop sideburns (which hid the scar where a pornographic publisher had slashed him with a knife) made him a favorite of caricaturists. He proudly claimed to have driven abortionists, gamblers, and quacks to suicide. Unlike members of our modern morals police, who tend to hurl their sticks and stones from pulpits and opinion pages, Comstock carried an actual badge, granted to him as a special agent of the Post Office Department (which later became the U.S. Postal Service). Ironically, he had several enemies in common with Macfadden: alcohol, cigarettes, and patent medicines. But they also shared the blind righteousness of the crusader whose message has found eager listeners. On the issue of prudery it was inevitable that they would lock horns.

On October 6, the *Times* ran a second, very different, story about the exhibition. The multipart headline appeared on page 9:

COMSTOCK TAKES HAND IN PHYSICAL CULTURE SHOW
Has Promoters Arrested for Putting Up Posters
SHOWMAN SAYS HE'LL FIGHT
How Can Anyone Show How Physical Culture Has Improved the
Body If the Body Is Clothed?

On the night of October 5, just two days before the exhibition
was set to officially open, Comstock raided the offices of the
Physical Culture Publishing Company. He'd visited three judges
before finding one who'd grant him a search warrant, but once he
had one in hand he sped to Nineteenth Street, accompanied by a
half-dozen vice squad officers. He encountered Macfadden at his
desk, dressed in an ensemble of "sandals, a broad-brimmed hat
and other clothes as well." Comstock ordered Macfadden placed
under arrest, on the charge of circulating obscene pictures,
namely, the posters featuring photos of the previous contest's
female winners. Treloar's loincloth didn't seem to bother him.
Comstock claimed that the posters proved the exhibition "proposed to exhibit the forms of young women, denuded of their
proper womanly apparel, for young men and others to look at,
and pay." Macfadden said that Comstock had also been enraged
by a photograph on the wall of the topless—and armless—Venus
de Milo.

Macfadden was defiant. "The purpose of this exhibition,
which has now become annual, is to show how the spread of
physical culture has improved the human body," he told a reporter after being released on $1,000 bail. "Manifestly that can-
not be done if the exhibitors are covered with clothing. All of this
was done last year and no one became indignant." Macfadden was
either blissfully unaware of how his exhibition was viewed by the
general public or had his tongue firmly planted in his well-exercised
cheek as he took the moral high ground; *Physical Culture* was
already notorious for publishing photos of topless women exercising, and running sensational articles with titles like "Naked
People Near New York," which gushed about a nudist colony in

New Jersey. Comstock seized five hundred pounds of posters. Macfadden promised, "I'll fight this case in every court in the land before I'll let it drop."

At this point, what had been shaping up to be a popular show became a national sensation. The next day's *Times* published an editorial titled "Comstockery Vs. Macfaddenism," which allowed that the promoter was genuine in his desire to show off "the culling of the fruits of physical culture," but sided with Comstock by saying that such an exhibition should not be "a money-making show for the express attraction of the 'baser nature,' which will be its effect if not his purpose."

Two more days of newspaper coverage followed, and on opening night twenty thousand people showed up at the Garden, five thousand more than could be squeezed inside. By 8:45 p.m., the New York City fire inspector had slammed shut the gates. Only twice before in its history had the arena been forced to turn away paying customers.

To Macfadden, the crowds were evidence that "it cannot be questioned that the American people are awakening to the benefits of Physical Culture." The week's drama had also provided him with something that always spurred him to new heights: an enemy. Though he was scheduled to go to trial in early 1906, Macfadden could not resist taunting his new foil in the pages of his magazine. The scathing lead editorial in the December 1905 issue of *Physical Culture*, the first to be sent to the printer after the exhibition, was titled "Comstock, King of the Prudes."

He thinks he is working in a good cause. He does not realize that by branding the human body as a thing obscene he is casting an insult at the great creator. . . . He is "protecting" boys and girls by putting up bulwarks, by blindfolding them that they may drift unknowingly into gross and vile habits that sap and destroy manhood and womanhood before the latter have even had a chance to develop into bud. . . . THE MOST INFERNAL ENEMY OF THE BOYS AND GIRLS OF THIS COUNTRY, AND OF ALL COUNTRIES WHERE

THE ENGLISH LANGUAGE IS SPOKEN AND READ, IS COMSTOCK. HE IS THE QUINTESSENCE OF PRUDERY.

On March 28, 1906, Macfadden was found guilty but released on his own recognizance. Both sides claimed victory. Macfadden would continue to ridicule Comstock in editorials up to and long after his foe's death in 1915.

By the spring of 1906, Macfadden's superhuman energies were already focused on a new project across the Hudson River, where his boldest scheme yet was coming to fruition. But the tussle with Comstock had pegged him as a troublemaker, and would start a process that would nearly destroy everything he had built.

Chapter 5

Utopia, New Jersey

*We would all live together as nearly naked as possible, for that
was, after all, the only pure thing. . . . We would make our
livings by the manufacture of all sorts of exciting apparatus and
health foods. . . . Men and women would wander forth from
the great, unclean, insanitary cities and live in clusters of pretty
cottages. . . . And the human body would become holy.*

—HARRY KEMP, FROM *TRAMPING ON LIFE*, A NOVEL BASED
ON HIS EXPERIENCES AT A CERTAIN HEALTH REFORMER'S
UTOPIAN COMMUNITY

The first decade of the twentieth century was a high-water
mark for the Progressive Era. In the thirty years between
1870 and 1900, Americans abandoned their farms for the industrial cities, and the country's urban population swelled from just
under 10 million to more than 30 million. "It became increasingly evident that all this material growth had been achieved at
a terrible cost in human values and in the waste of natural
resources," wrote Richard Hoftstadter in *The Age of Reform*.
Women's rights, safe working conditions, child labor, rapacious
monopolists, and overcrowded tenements were among the many
topics of debate popular in reform circles. Muckraking journalists were exposing the evils of modern society. The very week

that Macfadden was squaring off against Comstock over the merits of his New York exhibition, *Collier's* magazine scored a powerful, and eventually fatal, blow against one of Macfadden's sworn enemies, the patent medicine industry—makers of unregulated nostrums and elixirs that are today associated with medicine shows and snake-oil salesmen but in 1905 were a serious, and seriously profitable, business. In the first decade of the twentieth century, there was no shortage of people who thought that if society could weed out evil, human potential was unlimited.

No voice in the progressive choir sang louder for the improvement of the American people than Bernarr Macfadden's. For several months in 1902 and 1903, he published a magazine variously titled *Macfadden's Weekly*, *The Cry for Justice*, and *Fair Play*. The journal billed itself as "a weekly publication devoted to new thought." Its populist party line called for smashing business monopolies, providing good public education, and demanding workers' rights. "IN ORGANIZATION THERE IS STRENGTH," Macfadden thundered in one editorial. "IN UNIONISM THERE IS POWER. LABOR IS THE REAL KING, THE REAL FORCE THAT MOVES THE CIVILIZED WORLD."

For the August 1903 issue of *Physical Culture*, Macfadden penned an editorial that outlined what he called a "dream": a healthy, self-contained utopian community founded on the principles of clean living and plentiful exercise. For a dream, his vision was awfully specific—the settlement would be built on at least a thousand acres of land and located within an hour's journey of New York and Philadelphia. "Imagine, if you can, what an object lesson to this country, even to the world, a city like this could be," he wrote.

At the turn of the century, the golden age of American utopian communities had just about reached its sell-by date. The nineteenth century had been a fertile time for do-it-yourself Edens, including the Amana Colonies in Iowa, the Oneida Community in New York, and the twenty or so Shaker settlements scattered across the country. One remaining community, frequently praised in the pages of *Physical Culture*, was Zion City. Its founder was John Alexander Dowie, a charismatic preacher who gathered five thousand followers north of Chicago at the tail end of the nineteenth century. Zion City was a deeply Christian development, where the long list of taboos included smoking, drinking, pork, apothecaries, and especially vaccination, which is probably why Macfadden hailed Dowie in 1903 as "the greatest leader and most remarkable reformer in a century." Dowie slowly ceased to differentiate between himself and the Almighty and was eventually suspended from his church for financial improprieties, but Macfadden did not forsake him. As someone who had achieved the dream of building a utopia for refugees from the plague of inoculation, Dowie's credentials were unimpeachable.

That did not mean his work couldn't be improved upon. The name that Macfadden chose for his dream town expressed his ambitions for the settlement: Physical Culture City. In 1904, Macfadden dropped hints in *Physical Culture* that he might be looking for a large, cheap chunk of land. Early the next year, the *Janesburg Record* reported that "the well-known professor of physical culture [has] purchased 1800 acres from the estate of a local family near Spottswood," in the northern part of New Jersey. The parcel was well suited to his plans. The five-mile-by-two-mile size allowed for hundreds of lots to be divided. The location was forty miles from New York City and fifty from Philadelphia, and was bisected by the Pennsylvania Railroad, an invaluable link in his plan to shift the bulk of *Physical Culture*'s offices and printing facilities to their new home. Physical Culture City even had its own seventy-acre body of water, which the proprietor proudly dubbed Lake Marguerite.

The name was a tribute to his new wife. Marguerite Macfadden, née Kelly, was a Canadian nurse whom Macfadden married in Hoboken, New Jersey, in 1903. A Macfadden Publications employee once noted that her boss's tastes in women ran toward "Amazons" with pretty faces, large breasts, and wide, strong hips that he associated with successful childbearing. Marguerite fit the profile perfectly. She almost immediately began contributing articles on raising children to *Physical Culture* and a sister magazine that he'd started, *Beauty and Health*—a sign that Macfadden was thinking about starting a family. The couple rented a small house twenty miles outside of New York City, where they spent their first months of wedded bliss living the chapters of Macfadden's latest book (his tenth in five years), *Marriage: A Lifelong Honeymoon*, in which its author posited that the causes of divorce were "all largely physical," a result of being out of shape. The vigorous newlyweds worked the land, slept with the windows open, and prepared for the doctorless birth of their daughter Byrne. The proud papa published a photo of his seven-day-old little girl in the March 1905 issue of *Physical Culture*.

By June 1905, Macfadden had moved into a specially built $1,500 six-room home with thirty-eight windows for maximum sunlight and air flow. Physical Culture City was open for business, and its founding father was boasting of "a community [with] no sickly prudes, no saloons, drug stores, tobacco shops or places in which one may purchase things that make for the moral undoing of man or woman," whose citizenry was composed of "men and women whose alert minds shall be sheltered in healthy bodies." Red meat, high-heeled shoes, and white bread were also forbidden; violators of the ban risked expulsion. Building lots were relatively cheap—$15 for a 100' × 50' site and $50 and up for sites of an acre or more—but Macfadden insisted on granting ninety-nine-year leases rather than selling plots, to assure that his strictures would be adhered to.

Even had there been a pub on every corner, it's unlikely that the first settlers of Physical Culture City would have had many idle hours to pass at them. Or much money to do so. With the exception of a few wealthy early birds who came for a summer of spa-style detoxification, these first arrivals tended to be refugees from the nearby metropolises, young dreamers long on devotion to the ideas espoused in *Physical Culture* and short on cash. Shirtless men and bare-legged women (and the occasional nude of either sex) set to work clearing the Jersey pine forests, carving out streets, and erecting the simple cabins that would be their new homes. Laborers earned a dollar a day plus room and board. Many of these were students who applied most of their wages toward an advanced degree in physical culture that Macfadden was offering, and were taking home about fifteen cents a day after fees were deducted. The program of study was scheduled to run five and a half years. "It was suggested by some of the less charitable students that by that time Macfadden would have an attractively developed piece of real estate," noted one observer.

Among the pioneers was Harry Kemp, a well-known "tramp poet" of the early twentieth century who is remembered less today for his mediocre verse than for inspiring Jack Kerouac to go on the road, and for running off with the wife of his fellow writer and physical culturist Upton Sinclair. Kemp was a regular reader of *Physical Culture* who had come to Physical Culture City to live according to the master's teachings. In his lightly fictionalized memoir *Tramping on Life*, he recalled the bliss of building Macfadden's corporeal capital, living on one bowl of soaked wheat kernels a day, and absorbing the sun's rays on his naked body. Among the faddists he rubbed bare elbows with were a man who found relaxation by burying himself up to the neck in mud; "a little brown woman like the shrivelled inside of a walnut," who refused to imbibe any fluids except those she absorbed through eating fresh fruit; and a pair of young Italians who ate nothing but uncooked fruits and vegetables. "If appearances prove anything, theirs was the theory nearest right," Kemp wrote of the raw foodists. "They were like two fine, sleek animals. A fire of health shone in their eyes."*

Two hundred pilgrims arrived that first summer. All seemed

---

*A hundred years later, raw foodists still insist their eating program gives followers "the glow."

to thrive in the unorthodox setting, improving their bodies—an early morning bell summoned residents for a calisthenics class that Macfadden himself often attended—and pushing the boundaries of acceptable attire. At Lake Marguerite, women donned skirtless one-piece men's bathing suits, men wore only trunks, and children as often as not skinny-dipped. Kemp swam in a G-string until Macfadden pulled him aside one day and told the poet that while *he* personally didn't care what Kemp wore, some other bathers objected.

Macfadden may also have had in mind the sensibilities of his neighbors. Physical Culture City had quickly become a hot topic of gossip in Middlesex County, and Kemp recalled residents from the surrounding area converging on the settlement each Sunday to watch the goings-on. "They do not like our apparel," Macfadden fumed in a *Physical Culture* editorial after that first summer, noting that he was "fully aware that many people think that we are a lot of crazy fanatics." Passengers on the Pennsylvania Railroad crowded train windows and poked their heads out to get a good view of the skin show.

With rumors of a nonstop vegetarian orgy swirling across the Hudson River, it was only a matter of time before one of the New York papers picked up the scent of a juicy story. The colorful *New York World* dispatched a reporter, and he came back with a doozy of a tale. Describing Macfadden's advantageous earn-while-you-learn system, he wrote, "It was all work and no study. . . . The contract provisions were so arranged that graduation was virtually unobtainable." The writer went on to imply that "the researches among the boy and girl students were indeed exhaustive." Macfadden was furious, but rather than wait for the story to blow over, he went on the offensive, suing the *World* for $50,000 in libel damages. A jury eventually sided with the *World*, concluding that the reporter had only chronicled what he'd witnessed.

Macfadden had bigger headaches to contend with. *Physical Culture*'s editorial offices had been transplanted to the magazine's namesake city, but not everyone was equally enthused about its arrival. The Pennsylvania Railroad had at first responded eagerly to Macfadden's inquiries about having a station built, until he told them he wanted the name Physical Culture City painted on the depot's sign. The building was eventually constructed, but the depot remained anonymous.

The local postal service was even more excited to do business with the publisher, who would be spending almost $15,000 a year mailing his magazines and books, a spectacular sum in rural New Jersey. The postmasters of both Spottswood and its neighbor Helmetta wanted this business, and Spottswood landed it. The Spottswood postmaster earned a huge raise. Macfadden earned the lasting spite of Helmetta's town fathers. Underdressed Physical Culture City residents were quickly forbidden from jogging on the town's freshly paved roads; those who walked into downtown Helmetta wearing bathing suits were arrested.

Macfadden's legendary persuasiveness would probably have smoothed things over eventually if his legendary pugnaciousness hadn't already irked Anthony Comstock and his friends at the Post Office Department. The *Phrenological Journal*, the country's leading scholarly publication of personality evaluations based on the shape of a subject's head, astutely noted in 1908 that Macfadden "receives strong impressions, and strikes when the impression is hot, and before his ardor has cooled."

Prudery, had, of course, topped Macfadden's list of "the horrible curses of humanity." A *Physical Culture* reader wrote in to the editor's monthly General Question Department inquiring as to "the exact meaning of the term," and Macfadden obliged with an explanation. Prudes, he wrote, were people who ran around shouting "Shame! Shame!" any time the subject of procreation was raised, and "Would have us believe that ignorance to sex means innocence and virtue . . . they have placed upon the reproductive function the stigma of impurity." The reality, as Macfadden saw it, was that everyone was going to be faced with "sexual quick-sands and pitfalls" at some point in their lives, and only with a basic knowledge of the facts of life would they be able to sidestep these perils. Since sex education was virtually unknown, he volunteered to serve as America's guide.

The twin traps to which Macfadden refers were syphilis and gonorrhea, and it is difficult to imagine in the era of penicillin the degree to which these diseases ran rampant in the first decade of the twentieth century. In 1909, one out of five Army soldiers was estimated to have a venereal disease. That figure may have been low because both syphilis and gonorrhea have long dormant periods and only those soldiers whose symptoms were manifest were counted. Because sex was not an appropriate topic for discussion or magazine stories in the post-Victorian era, VD wasn't much talked about either. Grooms were generally not expected to inform their future mate of such an infection; that a bride-to-be would have caught such a bug was unthinkable. If a man with VD infected his wife and they had a child, they ran the risk of birth defects (blindness, especially), and sterility for the mother. The bacterium would be passed on to the next generation, continuing the cycle.

In 1906, newspapers still largely avoided the subject, except in the patent medicine advertisements that they accepted for venereal "cures."* Prince Morrow, a pioneering MD in the study of VD, noted that, "Social sentiment holds that it is a greater violation of the properties of life publicly to mention venereal disease than privately to contract it."

"Growing to Manhood in Civilized (?) Society," which was

---

*The most common "cure" for syphilis was dosing a patient with mercury, either orally, topically, or by inhalation. In his book No Magic Bullet: A Social History of Venereal Disease in the United States Since 1880, Allan M. Brandt describes one method of treatment for chordee, a "curvature of the penis which caused pain upon erection" which can result from gonorrhea. "[The doctor] recommended that the organ be placed 'with the curve upward on a table and struck a violent blow with a book . . . and so flattening it.'"

published in the October 1906 issue of *Physical Culture*, purported to be the confessions of sixteen-year-old rake Reginald Piel Barnes-Carter. It actually sprang from the pen of Macfadden's ace writer, John Coryell, whose protagonist voices some of his boss's strongest feelings about prudery. Because Barnes-Carter, like so many young men and women of his time, is kept ignorant of the facts of life, he soon falls victim to syphilis, which at that time was still a likely sentence of insanity and early death. He goes on to seduce, and infect, two girls before he learns of his disease, at which point he rails against prudery and "society's determination to foster hypocrisy on sex matters."

Macfadden's own sexual habits were causing a bit of a stir within the Physical Culture City community. His lifelong honeymoon with Marguerite seems to have been derailed shortly after Byrne's birth. Macfadden was spotted on more than one occasion walking hand-in-hand with his secretary, Susie Wood, a large-boned girl who worshipped Macfadden as a health genius. (Macfadden's third wife, Mary, would say of Wood's adulation that "she thought he was a sort of Darwin of a new way of life.") Eventually, the two set up house in Macfadden's window-saturated home overlooking the lake named for his wife, with Susie's mother serving "as ostensible housekeeper," according to one Physical Culture City resident. Marguerite returned to Canada with baby Byrne and disappeared from Macfadden's life. Susie became pregnant and delivered a baby girl in 1906.

Though Macfadden had expressed a wish "to see Physical Culture City gradually developed into an industrial city of size and importance and wealth," the population of his whole-grain Athens stagnated at a village-like two hundred. When word of his shenanigans inevitably leaked out to the faithful who'd listened to his lectures on the sanctity of marriage, even that number began to dwindle. By March 1907, they had another reason to leave. Macfadden had been arrested again, this time for sending what the complaint identified as "obscene, lewd, and lascivious" materials through the mail—namely, the "Growing to Manhood" serial. The arrest stemmed from an anonymous tip from one of the residents of Spottswood. Macfadden was never able to unmask his accuser, but he constructed a plausible theory. He had just received approval from the federal postal service to open a Physical Culture City post office, which would have meant a severe reduction in the standard of living of Spottswood's postmaster. As for the smut in question, Macfadden wrote in 1934, the same year that Henry Miller's *Tropic of Cancer* first appeared, that the serial "could be published at this time and it would be a child's primer in comparison to some of the books that are now freely distributed everywhere." At the time, however, the court found such steamy prose as "I took her on my lap and kissed and fondled her" too hot to handle. The judge ordered parts of the story excised from the court record.

Again, Macfadden swore to fight his case to the death. He orchestrated a massive letter-writing campaign to President Theodore Roosevelt, who had received glowing press in *Physical Culture* for his advocacy of a "strenuous life" of exercise in the outdoors. Roosevelt ignored his plea. As Macfadden filed his appeal of the sentence, his utopia began to crumble. A last-ditch town meeting ended with Macfadden losing his temper and accusing several residents of physical-culture treason. By July 1907, the offices of the Physical Culture Publishing Company had moved from the wilds of New Jersey to the ultramodern Flatiron Building in New York City.

Macfadden arrived at the Federal Court House in Trenton, New Jersey, on October 22, 1907, to hear his fate. He did not have to wait long. It took the jury just a few minutes to find him guilty of "mailing non-mailable materials." His sentence: two years of hard labor in the New Jersey Penitentiary and a $2,000 fine. After an eighteen-month battle, the Supreme Court declined to hear his appeal. He was spared the agony of prison when President William H. Taft—who, oddly enough, was famous for being so fat he'd gotten stuck in his own bathtub—commuted his sentence. Macfadden grudgingly paid the fine. He spent the rest of his life fighting to get his $2,000 back.

By this time, Physical Culture City was a ghost town. Aside from the original grid of streets laid out by Macfadden's exhibitionist students, few traces of it remain today in the Jersey exurbs; Lake Marguerite was drained, and the disputed train depot is long gone. After losing an estimated $100,000 trying to build his Eden, Macfadden the dreamy progressive was beginning to give way to Macfadden the pragmatic businessman. "Colonies based on ideals seem doomed in advance to failure in nearly every instance," he later wrote. "And when greed and human selfishness assume a weighty influence within their midst they can hardly be expected to survive." Sensing opportunity in a crisis, he rounded up his remaining faithful and shipped them off via railroad to Michigan.

Health Guru

—from the twenty-five-page index of specialized treatments
in *Macfadden's Encyclopedia of Physical Culture*

Battle Creek, Michigan, was to American health and fitness in 1907 what Las Vegas is to gambling today. Trainloads of overfed and stressed-out pilgrims alighted there from all around the country, most of them headed for the world-famous Battle Creek Sanitarium. The popularity of this enormous facility—usually called simply the San—and the ancillary health businesses that sprang up around it was due to the work of its benevolent dictator, John Harvey Kellogg. More than three hundred thousand health seekers would pass through the doors of the San during its sixty-five years of operation under Kellogg, most of them drawn to some degree by the doctor's reputation as a gifted all-natural healer who understood that his guests wanted to suffer their privations in comfort. Kellogg himself had arrived when summoned, indirectly at least, by a higher power—God.

The conduit for the Deity's job offer was Ellen White, high priestess of the Seventh Day Adventist church. White, whose religious visions formed the backbone of church doctrine, found herself taking an urgent health-related memo from the Lord on the evening of Friday, June 6, 1863. From that day forward, He decreed, Adventists were to eat two meals a day, avoid meat (which brought forth animal desires and a thirst for whiskey), consume whole-wheat bread, drink only water, and shun all sweets, lard, and spices. The Deity, skeptics noted, seemed to be remarkably well read in contemporary health-reform ideas, particularly those of James Caleb Jackson, whose sanitarium in Dansville, New York, White had recently patronized. White chalked up any similarities to coincidence.

A second revelation on Christmas Day, 1865, informed White that the church was to move to Battle Creek and found a convalescent home. The Western Reform Health Institute opened in 1866 and was an immediate flop. White's solution was to hand-pick Kellogg, one of the church's most promising youngsters, and pay to send him off to earn his medical degree. When he returned from New York City in 1876 to take over the San, John Harvey Kellogg was a gifted surgeon. His marketing skills weren't lackluster, either. In 1876, the San treated twelve patients. Six years later, it was advertising itself as the largest facility of its kind "in the West." Kellogg would become the world's best-known vegetarian and colon-flusher.*

---

*T. Coraghessan Boyle's historical novel The Road to Wellville—the title was taken from a C. W. Post pamphlet of the same name—is set at the San in 1907 and has Kellogg as a main character who considers his young rival, Macfadden, to be "a harebrained, posturing, bare-chested, dumbbell-thumping parody of a health professional." Boyle's rendering of a Kellogg speech offers a glimpse into the recreational pleasures a Battle Creek vacationer might have experienced: "Ah, the exquisite and unremitting agonies of

Directly across the street from Kellogg's San sat a five-hundred-room brick building owned by his chief rival, the cereal magnate C. W. Post. Post, who'd once been a guest of the San, and had stolen from it (in Kellogg's version of the story, anyway) the idea for a grain-based coffee substitute. Sold as Postum, the beverage made Post rich. He used this money to launch Grape-Nuts, thus altering breakfast history. In response to Post's success, John Harvey's younger brother Will Kellogg founded the Kellogg Cereal Company, which still operates out of Battle Creek. It was this Kellogg who brought the world cornflakes, as well as Sugar Smacks, Pop Tarts, and dozens of other highly processed food products that surely taunt John Harvey's ghost. The Kellogg brothers parted ways after Will's success, and in 1907 Ellen White excommunicated both of them from the Adventist church for having strayed too far from her teachings.

Because two breakfast-cereal millionaires had been created there almost overnight, trainloads of health-food entrepreneurs disembarked at the Battle Creek depot. Among them was Macfadden, who had devised a breakfast cereal of his own: Strengtho.

C. W. Post's building had originally been constructed by a team of entrepreneurs as the Phelps Medical and Surgical Sanatorium. Its backers had planned to steal customers from the San by serving meat and allowing smoking. Within four years the Phelps was bankrupt. Post picked it up at auction and not long after rented it to Macfadden, who Post must have suspected would drive John Harvey Kellogg bananas.

By the middle of 1907, Macfadden had installed his transplanted students from Physical Culture City. When the doors of the Bernarr Macfadden Sanatorium at Battle Creek were opened, the contrast with the austere facilities at Physical Culture City could not have been more striking. This wasn't a place to pitch

---

the flesh eater, his colon clogged with its putrefactive load, the blood settling in his gut, the carnivore's rage building in his brittle heart—a steak kills day by day, minute by minute, through the martyrdom of a lifetime."

a tent or chop wood, but a "mansion with a white columned rotunda, massive fireplace and large lounging chairs . . . gym, swimming pools, Russian and Turkish baths" and other unspecified "recreation facilities." Before long, Macfadden was describing it as the world's "largest and most magnificent institution devoted exclusively to the treatment of disease through natural methods"—a crunchier-than-thou dig at Kellogg, who, of course, was a surgeon as well as a holistic healer.

Here the onetime kinisitherapist's gift for nomenclature surfaced again, as he unveiled the combined therapy that he called "physcultopathy." (In print, he often preferred to capitalize its grander full name, "Physcultopathy: The New Science of Healing.") Whereas Kellogg marketed his San primarily as a therapeutic getaway, Macfadden saw his health home less as a resort destination than as a destination of last resort. In Battle Creek, Macfadden once again put out a call to the wretched souls whose ailments had baffled their physicians. "In nearly all cases," he wrote of the incurables who converged on his health facility, "the various methods advocated by medical science have been tried out thoroughly before the patient has turned to Physcultopathy."

The individual elements of The New Science of Healing weren't exactly novel—Macfadden admitted his debt to Sylvester Graham, Dudley Sargent, and others, and the core of his philosophy had been pretty much summed up by the title of his book *Fasting, Hydropathy and Exercise: Nature's Wonderful Remedies for the Cure of All Chronic and Acute Diseases.* The ways in which Macfadden *combined* his alternative therapies were original, however. As always, he advocated the use of "scientific feeding," a sliding scale of dietary abstinence calibrated to the particular needs of a patient, supplemented with exercise. But the home in Battle Creek also put out the welcome mat to practitioners of various forms of alternative medicine. His chief of medicine was a naturopath, a doctor who eschewed any nonnatural methods of healing. Osteopathy, the system of treating illness through skeletal manipulation that had emerged from Missouri around the same time as Macfadden, was offered on site, as was more traditional muscle massage. Macfadden was especially enthusiastic about his water cures. "Hydrotherapy is used to a considerable extent in nearly all cases," he wrote. "Turkish baths, Russian baths, plunge baths." If a treatment could be categorized as "natural," Macfadden would offer it to his clientele. If the American Medical Association didn't like it, all the better.

One lesson Macfadden had learned from Kellogg, as well as from his disasters at Lake Ronkonkoma and Physical Culture City, was that paying guests on holiday expected more for their money than to be roused every morning for a long hike and have their food supply cut off. The photographs that accompanied a story in *Physical Culture* about the "astounding cures effected" by physcultopathy took pains to showcase "the commodious kitchen" in which meals were prepared, and "the magnificent dining room, finished in mission style, decorated and furnished as well as some of the most elaborate and beautiful dining rooms in America's palatial hotels." The intent was to make patients feel like they were vacationing, and Macfadden at least thought that he succeeded. "On more than one occasion I have had patients say to me on leaving, 'I am not only pleased with the results of the treatment I have received here, I not only feel stronger and better in every way, *but I never had such a good time in all my life.*" For entertainment on Friday evenings, Macfadden gave lectures on topics such as "Medicine—the Science of Guessing," which he would punctuate with feats of strength, including his black-cabinet posing routine. As an encore, Macfadden would perch his feet on a table, extend his body and do twenty-five one-handed pushups, picking up a match in his teeth with each descent.

Among the most satisfied customers of the Macfadden Sanatorium was the muckraking novelist Upton Sinclair. The author of the stomach-turning slaughterhouse novel *The Jungle* was a newly minted literary and progressive superstar at the time he arrived for a dose of physcultopathy at the end of the summer of 1909. (Macfadden had excerpted *The Jungle* in 1906 and offered a special edition of the book as a *Physical Culture* subscription premium.) Sinclair was writing an occasional column for *Physical Culture* and arrived at the home exhausted from finishing his socialist novel *Samuel the Seeker*. He wanted to undergo one of Macfadden's fasting cures. Their friendship was immediate. Macfadden respected Sinclair for his ability to bring about change in Washington, D.C. Sinclair, who was one of the great health faddists of the twentieth century, seemed positively starstruck in the presence of B.M., as he called him. This was, in fact, Macfadden's preferred nickname, and the associations those initials had with the digestive process made him like it even more. Macfadden, Sinclair wrote two decades after this first meeting, taught him "free, gratis and for nothing, more about the principles of keeping well and fit for my work, than all the orthodox and ordained physicians who charged me many thousands of dollars for it."

Sinclair was among those whose ailments—headaches, indigestion, rapid weight loss—had baffled the medical establishment. He began his stay with a fast of ten to twelve days, and followed with one of Macfadden's new discoveries—an all-milk regimen. His description of his time at the Battle Creek resort is the finest that exists:

After the fast we went on a thing known as a "milk diet"; absorbing a glass of fresh milk every half hour, and sometimes every twenty minutes, until we got up to eight quarts a day. The fasters sat around, pale and feeble in the sunshine, while the milk-drinkers swarmed at the dairy-counter, and bloomed and expanded and swapped anecdotes—it was a laboratory of ideas, and if you had a new one, no matter how queer, you could find somebody who had tried it, or was ready to try it forthwith. When you came off the milk diet, you might try some odd combination such as sour milk and dates. In the big dining-room you were served every sort of vegetarian food—and there were dark rumors that the smell of beefsteaks came from Macfadden's private quarters. I asked him about it, and he told me he was trying another experiment.

The Macfadden Sanatorium was a modest success. Strengtho was not. Macfadden may have been correct in claiming that his cereal was nutritionally superior to Kellogg's Corn Flakes, but Kellogg definitely had a packaging advantage. Battle Creek's great breakfast cereal successes had relied on convenience, whereas Strengtho, because it was made with untreated wheat germ, needed to be cooked almost instantly after purchase or it went stale. As for the beefsteaks, Macfadden was never a strict vegetarian. He simply recommended vegetarianism for those whom he suspected lacked his dietary discipline. Considering that he often got by on one meal a day and never ate *anything* on Mondays, he may have been right.

When Macfadden's two-year lease was up, he decided not to exercise an option to buy. Post, he said, had been "skinning him on the rent." He also discovered that the purchase price he'd agreed to was significantly higher than what Post had asked from others. Once again, Macfadden began to search for new quarters. He found them on Chicago's South Side.

The Bernarr Macfadden Healthatorium, a massive Gothic-style building on Chicago's Grand Boulevard, was well suited to Macfadden's grandiose plans both in size and location: It stood five stories tall and occupied an entire block in what he called one of the city's "fancier precincts." It had until recently served as the headquarters of the Lakeside Club, a prominent Jewish social organization. The structure was valued at $250,000, but Macfadden boasted of having purchased and renovated it for less than half that amount. Grand Boulevard (which has since been renamed Martin Luther King Boulevard) had a wide grassy esplanade down its center, where Macfadden could enjoy his daily barefoot promenade with students in tow. The stench from the nearby Chicago stockyards that Upton Sinclair had made infamous in

The Jungle would have served as a powerful reminder to stick with the vegetarian program.

As usual, Macfadden was doing the work of several men while at the Healthatorium. He was editing *Physical Culture* and commuting frequently to the Macfadden Publishing business offices in New York City, and supplementing his income by giving speeches and churning out a new health book every few months. He had expanded his curriculum at the Physical Culture Training School, offering day and night classes toward a "Doctor of Physcultopathy" degree. He personally interviewed and devised a treatment plan for each of the thousands of patients who arrived for an extended stay, including one 300-pound Minnesota hotelkeeper whom Macfadden quickly slimmed down to 225 on a water-only fast. The Healthatorium's facilities, which dwarfed those at Battle Creek, included a full-size indoor swimming pool, and a 90' × 60' gymnasium with 35' ceilings. Macfadden crammed this room with state-of-the-art fitness equipment.

To assure that the gears of this physcultopathic machine ran smoothly as he worked on his next project, Macfadden appointed a manager whose loyalty and efficiency could not be questioned: Susie Wood, his housemate from Physical Culture City. Wood brought along their daughter and her mousy new husband, Morey. Macfadden needed as much managerial assistance as he could get, since he was also in the final stages of overseeing the writing, editing, and illustration of *Macfadden's Encyclopedia of Physical Culture*, a five-volume set of reference books that aspired to be the *Britannica* of natural health. The subtitle on the title page of the *Encyclopedia* announces its modest intentions: *A Work of Reference, Providing Complete Instructions for the Cure of All Diseases through Physcultopathy, with General Information on Natural Methods of Health-Building and a Description of the Anatomy and Physiology of the Human Body.*

Macfadden set out in his preface to dispel any doubts about

his credentials. "I am not dealing with mere theories in this work," he wrote. "I am dealing with facts that have been definitively proven in thousands of cases," those being his pupils, patients, and the readers of *Physical Culture* and his dozens of books. In addition, he noted, he had scoured the works of the world's great thinkers to assemble the best of what had been thought and said on the topic of natural healing methods. "The healthy person . . . is a radiating center of life—physical, mental, spiritual," he wrote. "To grasp his hand is a pleasure, to gaze into his eyes a joy."

The thousands of topics covered by the *Encyclopedia* sprawled over 2,966 pages. For 1912—a year in which the first vitamin was identified and American women were still banned from competing in any Olympic event that did not require them to wear long skirts—it was an astoundingly forward-thinking work. In just the first volume, Macfadden explained that he believed that the average woman should be as strong as the average man, applauded Indian yogis for their work on deep breathing, and recommended olive oil for cooking instead of lard. He offered recipes for meat-free menus, many with unappetizing photographs; the "vegetable turkey" looked more like a sun-baked cow pie. He boasted that only whole-grain loaves were served at his Physical Culture Restaurants, and he scoffed at the notion that white-flour bread could be called the staff of life. "It is more like the staff of death," he wrote.

The second volume gave descriptions of almost every manner of sport and exercise imaginable, with special emphasis on how to build a strong body for the purpose of staying healthy. (One section was titled "Weight Lifting as an Exercise and not as a Means of Breaking World Strength Records.") Many of the routines were accompanied by nude photos. A foldout poster of Macfadden, dressed only in a pair of short-shorts, illustrated dozens of strength exercises that one can still see performed daily in any American gym. The Harlem Renaissance poet Jean Toomer had a typical life-altering reaction. "I bought [Macfadden's] encyclopedia, and, in addition to following his prescriptions, I began talking and arguing his ideas with everyone," he scribbled in his diary in 1916. "No-one knew the fight I was making. But, in time, every-one saw the results."

The second half of volume 2 introduced what Macfadden calls physcultopathy's "two most important ideas" about disease. "These are first, that in the main, *there is but one disease*, and second, that *disease is a beneficent process of nature instead of the enemy it seems to be*." As for the single-disease theory, Mac-fadden was referring to his belief that almost all maladies are caused by "impurity of the blood," a toxified state reached by the overconsumption of impure or unnatural food and drink or in-gestion of other synthetic poisons. White flour, processed foods, alcohol, and coffee fell into that category, as did any and all medicines. Macfadden didn't deny that germs existed or that they could cause trouble. He simply refused to admit that they were a threat to healthy persons. Macfadden declared that physcultopa-thy's diet and exercise program was a near-perfect deterrent against germ-borne illnesses, and insisted that if one *were* to be-come sick, almost any illness could be alleviated by natural meth-ods. His apothecary consisted mostly of fasting, fresh air, light exercise, and the absolute avoidance of all drugs, which, of course, would cause impure blood. "The healthy, vigorous, ath-letic student here has no fear of disease; he is immune. Uncon-sciously he lives in the recognition of a fact that the doctors and learned men have forgotten, viz., that disease can find no lodge-ment in a healthy body. One might swallow 'disease germs' by the million, but when a radiant, abounding health exists they would die of inanition and be expelled before they could do the slightest harm." Vaccination, or as Macfadden saw it, the unnec-essary pumping of dead germs into the bloodstream, was lunacy.

Volume 3 of the *Encyclopedia* explored what decades later came to be known as alternative therapies, including various types of hydrotherapy. Macfadden wrote that chiropractic was a useful tool in alleviating many maladies, although "to assume that all disorders are the results of spinal misplacements is illogical." Osteopaths had a better understanding of the spine, Macfadden wrote. Neither method was sufficient in his eyes, which is why he had gone one step further and developed his own "Mechanical Physcultopathic Treatment," which was superior because it concentrated on the ligaments and muscles around the vertebrae as much as on the backbone itself (it was rather like a combination of osteopathy and Pilates). The volume closed out with no fewer than 118 specialized fasts and dietetic regimens, each targeting a specific result. Among them were 8 raw-food diets and 43 milk-based ones.*

The fourth volume offered cures for everything from bubonic plague (enemas, hot and cold packs, fasting, and fresh air) to masturbation (sex education and more fresh air) to nightmares ("in nearly all cases the cause of this ailment is eating beyond the digestive capacity"). Overeating seemed to be the cause of most ailments, fasting the cure. In Macfadden's world, one not only starved a fever, but also a cold, a cough, hiccups, psoriasis, cancer, and just about everything else.

The most curious and, for its time, radical of the five volumes was the last, which dealt with relations between the sexes. "There is nothing more deplorably true than that the majority of even those who are married are not conversant with the very first and most fundamental of the physiological laws of sex," he wrote. Since prudery fed on ignorance, Macfadden set out to demystify the procreative act. He explained the birds and the bees and described the processes of pregnancy and birth. He provided

---

*Most of these diets, which vary mainly in the amounts of dairy consumed and the means by which it was to be ingested, were designed to get as much milk into a human being as he could swallow, as quickly as possible. "The purpose is noticeably to stimulate the circulation and prompt the growth of new cells, new tissues, and at the same time cause the prompt elimination of all waste and effete matter in the system," Macfadden wrote. "If you have not fasted over two or three days you can usually begin on [an eight-ounce] glass of milk every half hour the first day."

full-color diagrams of the male and female reproductive systems. This fifth volume wasn't a sex manual in the multiorgasmic modern sense. Macfadden's how-to advice didn't extend much beyond tips such as, "The average man would do well not to exceed [intercourse] three or four times a month." He ranted for several pages about the hypocrisy of American attitudes toward venereal disease. Most of all, he tried to make clear that he saw sex as a part of marriage, and that the primary purpose of sex is not pleasure or romantic love, but to achieve something higher:

Eugenics is the name of the modern study or science which is concerned with racial improvement through the better breeding of human beings, this is to be accomplished partly by encouraging the marriage of those who are normal and sound, in every way, but especially by discouraging and preventing the reproduction of those who are unfit for parenthood. . . . The consistent physical culturist, in the broad sense of the term as it applies to racial improvement, will therefore naturally concern himself with putting into practice the fundamental principles of this new science.

Macfadden was hardly alone in his desire to improve the human race through selective breeding. Woodrow Wilson and the civil rights pioneer W. E. B. DuBois were eugenics boosters as well. Where others saw an interesting sociological theory, however, Macfadden saw an urgent uncompleted task at the top of his to-do list.

The year 1912 should have been one of triumph for Macfadden. The first volume of the *Encyclopedia* had rolled off the printing press. After four failed attempts at establishing a world health headquarters, his Healthatorium was thriving under the sharp eye of Susie Wood. Corset manufacturers and patent-medicine fraudsters, two of his primary targets, had been all but eliminated.

And the ideas espoused in *Physical Culture* were indeed flowing into the mainstream. Macfadden now operated twenty health-food restaurants in cities throughout the Northeast; alternative therapies were growing so quickly that a National Association of Drugless Healers was founded that year; Upton Sinclair, arguably the most famous writer in America, had published a book titled *The Fasting Cure* and dedicated it to his good friend B.M.; and YMCA attendance had tripled in the first ten years of *Physical Culture's* existence. "The greatest impress, perhaps, that the physical culture movement has made on modern life has been the utter rout to which it has put the opponents of exercise as a means toward health," Macfadden wrote in the April 1912 issue.

Behind his usual bravado, however, Macfadden was nervous. Despite his presidential pardon, his legal defense bills had cost him tens of thousands of dollars, and his financial situation was precarious. Were it not for the income he made from lecturing and selling books through advertisements in *Physical Culture*, he might have gone broke. In August 1911, his estranged wife, Marguerite, had resurfaced to file for what was shaping up to be a very bitter divorce. WIFE WANTS DIVORCE FROM VEGETARIAN STRONGMAN FOR DESERTION read a headline in the *Chicago Tribune*. "She waited six years for her husband to come back," the accompanying story said. "But . . . he moved to Chicago instead."

In 1912, another ghost reappeared: Anthony Comstock. Macfadden's mockery notwithstanding, the Society for the Suppression of Vice was going strong, and Comstock's close relationship with the postal service had allowed the organization—under the 1873 Comstock Law, which banned the circulation of "lewd, obscene and lascivious" material through the mails—to crack down on any literature viewed as inappropriate. In early 1912, Macfadden's books were summarily banned from the mails. On June 13, just a few minutes before he was scheduled to deliver a lecture on "the development of beauty and charm in womanhood" before what the *Washington Post* called "a fashionable audience of three hundred women" in Washington, D.C., city police arrested him. The U.S. Attorney's office had requested the arrest, tipped off by "post office authorities." The *Post*'s account noted that Macfadden expressed "great indignation" over the action, and "declared that since he began his work a number of years ago he has been attacked by prudish people who do not like the truth told in public, and that his arrest was a result of this opposition." He vowed to continue his campaign, but the message sent by the postal authorities was clear: *Physical Culture*, which relied heavily on mail subscriptions, would be the next target.

Abruptly, Macfadden decided to leave the country. The editor's note in the August edition of *Physical Culture*, which would have been near completion in mid-June, was evidently torn up and replaced with one that began: "After this issue, PHYSICAL CULTURE begins a new era. I have arranged to relinquish active business and editorial control of the magazine."

Macfadden's stated purpose for handing over the reins to his associate editor, John Brennan, was that he was "heartily sick" of business. His aim for the coming months, he wrote, was to energize the Physical Culture Union, a grassroots organization that he vowed to build into "a mighty political power." Macfadden's complaints may have been honest, for he had been running his *Physical Culture* empire for fourteen years. He neglected to inform readers, however, that he had quietly severed his business ties with the Macfadden Healthatorium and transferred his stock holdings in the Physical Culture Publishing Company to his treasurer, Charles Desgrey, in anticipation of taking a long trip abroad. Desgrey was also named president of the company. The two had a handshake agreement that the shares and title would revert to their proper owner when Macfadden returned to the States. Desgrey would retain a portion of the company as a reward. In the interim, any dividends would be paid to Susie Wood at the Healthatorium.

In the autumn of 1912, the forty-four-year-old Bernarr Macfadden quietly slipped into Canada, and then once again sailed for Britain, a land he'd always felt appreciated him better than his own. He had brought copies of his banned books to sell and had sketched out a lecture tour. He also had a new idea for a contest. The grand prize, unadvertised, was a chance to become the third Mrs. Macfadden.

Chapter 7

Mary

Man is an organism—an animal. . . . And the laws of the
improvement of corn and race horses hold true for him also.

—CHARLES BENEDICT DAVENPORT, FROM HIS INFLUENTIAL 1911
BOOK, *HEREDITY IN RELATION TO GENETICS*

The issues of *Physical Culture* published in 1913 were filled with stories about selecting an appropriate mate—a sure sign that the topic was on the mind of the editor in exile. Early in that year, Milo Hastings, the magazine's house nutritionist and sometime stenographer for Macfadden's more radical ideas, penned a feature titled "Can We Breed Better Men?" Through Hastings, Macfadden's growing fascination with fashionable mating theories was expressed with typical *Physical Culture* subtlety. "Eugenics is the mightiest comet that ever came skidding into the little solar system of human thought," Hastings wrote. "Suppose we are breeding for 'a sound mind in a sound body' and have formulated a scheme of judging the applicants with a score card system not unlike that which they grade the Orpingtons at the County Fair. Here is 'A-1,' her score is ninety-five and three-quarters, the best applicant in the lot for the high and holy function of motherhood."

With a sudden abundance of time on his hands in the fall of

1912, Macfadden was applying himself to finding his own A-1, a physically superior female suitable for mating.

Upon his arrival in England, Macfadden opened a health home at Chesham, twenty-five miles north of his London offices, which were headquartered between the Strand and Fleet Street. He was publishing two magazines in Britain, *Physical Development* for men and *Beauty and Health* for women. Through the latter publication, he announced a nationwide contest to find "GREAT BRITAIN'S PERFECT WOMAN." The winner, selected on the basis of "health, muscular development and good looks" would receive a cash reward of £100.

Five hundred women from throughout the British Isles sent in photographs of themselves. All but one were pictured wearing tights, as the contest rules stipulated. Twenty-five finalists were invited to London for a closer inspection. Macfadden and his team of judges—including John Coryell, who since the "Growing to Manhood" incident had been banging out health stories with titles like "Honey and Its Food Value" under the pseudonym H. Mitchell Watchet until B.M. summoned him from New York specially for this urgent assignment—scrutinized their measurements.

Mary Williamson, nineteen, the daughter of a Yorkshire mechanical engineer, worked in a carpet mill in Halifax. Williamson was a swimming champion who had twice completed a fifteen-mile race down the Thames River. The sole entrant to defy contest rules, she had submitted a snapshot of herself attired not in tights but in a swimsuit cut with trunks that stopped six inches above the knee, which she later noted "did show off my figure." The dimensions of that figure promised an absolute athletic knockout, by Macfadden's standards: 38-25-39, with thighs twenty-four inches around. She was five-foot-five and 142 pounds, almost exactly matching Macfadden's measurements.

Upon receiving Mary's photo, Macfadden summoned her to London immediately. She first met the physical culturist on January 9, 1913. "He was elated to learn that I had never been vaccinated," she noted.

The judges didn't take long to reach a unanimous decision: Mary Williamson clearly was Great Britain's Perfect Woman. As part of her duties she was required to accompany Macfadden on a lecture tour throughout sixty cities and towns in England, Scotland, and Ireland. (A chaperone and tour manager joined them.) At the campaign's end, she would receive her £100 prize as well as a cut of profits from what she called "cheesecake" postcards of her in a swimsuit; her eye-popping measurements were listed on the flip side.

Macfadden's marketing instincts had been right again: The Perfect Woman contest had been a public-relations bonanza. Lectures sold out, and his self-help books were moving briskly. (The publicity was not uniformly positive; the London Daily Telegraph noted that "his voice sounded as if it came from a man talking through a bunghole with his head in a rain barrel.") Mary quickly grew accustomed to seeing her photograph inside the local paper of each town in which they stopped. Macfadden had polished his stage act to a high shine in the twenty years since he'd first seen Sandow flexing on the Chicago Midway, and crowds were thrilled by his black-cabinet poses and roars of "Weakness is a crime! Don't be a criminal!" He again called for volunteers from the audience and ridiculed their manhood in front of their wives, girlfriends, and Britain's Perfect Woman, challenging them to match him in deep knee bends until some men had to be sent to the hospital.

The climax of the act was new. Macfadden summoned his perfect woman, then proceeded to lie flat on his back, clad only in a minuscule breechclout. Fixed in the spotlight, Mary slowly climbed atop a high table to the strains of Chopin's Funeral March. Macfadden raised his hand and asked her to announce her exact weight. She then leapt off the table onto his stomach, to the gasps of the crowd. Each time, Macfadden jumped quickly to his feet and took a victory bow for physical culture.

The tour chugged along, with performances almost every

night. Mary was receiving mountains of fan mail, including not a few proposals of marriage. One came from a U.S. congressman who'd seen her photo in the *New York Herald*. Mary's chaperone noted that the arrival of these mash notes always left Macfadden in a sour mood.

One afternoon less than two months after they'd met, while on a twenty-mile hike through the countryside, Macfadden asked Mary to marry him. He'd fallen in love with her at first sight, he said, and had until meeting her despaired of ever starting the perfect physical-culture family. Mary, swept up by the romantic idea of Macfadden as a health hero leading wayward humanity back to the proper path, accepted. He stood on his head to celebrate and asked his betrothed to time him. He clocked in at a minute and four seconds.

Marriage to a health guru, Mary found, would entail more than vowing to honor and obey. Macfadden required his fiancée to sign an oath that she would raise their children according to the principles of physcultopathy, forswear all caffeinated beverages, and avoid doctors in sickness and in health—especially in sickness. He made it clear that he expected their many babies to be born without a physician present.

The pair took a break from the tour to return to London for a no-frills ceremony without witnesses, not even Mary's family. "The marriage was not announced," Mary observed icily forty years later.

On their wedding night, the Macfaddens exchanged gifts. Mary gave her forty-five-year-old groom a pair of pajamas, which he quickly cast aside on the grounds that sleepwear was a barrier to procreation. He gave his teenage bride an autographed copy of *The Athlete's Conquest*. "As Mrs. Macfadden, reading in bed was not to be one of my pastimes," she later wrote.

By midsummer 1913, the antipajama strategy had paid off. Mary was pregnant, and ecstatic. To her surprise, though, pregnancy had no effect on the lecture tour. She continued to leap onto her husband's stomach almost nightly. Macfadden insisted on keeping their marriage a secret, on the grounds that the Perfect Woman was a better draw if unattached, so she wore no wedding band. They kept separate rooms on the road, and Macfadden would slip into Mary's room to perform his husbandly duty. (The monthly limits on intercourse that he recommended in the *Encyclopedia* were for *average* men.) He put his wife on a regimen of two hundred deep knee bends a day and instituted a strict vegetarian diet.

"The children I was to have would be looked upon by Bernarr as the beginning of a new race of human beings uncontaminated by the things that had put civilization where he thought it was," Mary wrote. To her relief, the tour's momentum screeched to a halt in Scotland, after the *Glasgow Herald* cited a Macfadden article that equated whiskey drinking with quack medicine.

The Perfect Woman caravan finished with a July show at London's Royal Albert Hall, for which Macfadden insisted on having his four-months-pregnant wife leap from a height of five and a half feet. Afterward, the pair—to all appearances traveling like a father and daughter—moved south to Brighton, a resort town on England's southern coast. When Mary appeared on the cover of the August 16 edition of *My Weekly*, a popular women's magazine, the old wrestling promoter coaxed his wife into performing one last publicity stunt to promote the health home he planned to open. At summer's end, Mary, quite obviously with child, executed a perfect sixty-foot swan dive off the top platform of Brighton's Long West Pier in front of a cheering crowd.

The four-story Brighton home at No. 6 Eastern Terrace, which the Macfaddens rented starting August 1, had an impressive pedigree. It was the former residence of the late James Ashbury, a member of Parliament and yachting legend who'd vied several times for the America's Cup. General Ulysses S. Grant and his wife were Ashbury's guests at No. 6 after Grant's presidency.

Elder Brightonians recalled Grant's endorsement of the building's excellent wine cellar, which Mary noted was "now reduced to a repository for my husband's bottled beet juice, honey-water jars, carrot bins, boxes of nuts and macerated wheat, run-down *Physical Culture* watches [timepieces once given away to subscribers that listed the proper hours for one's twice-daily meals and exercise breaks], discouraged punching bags [and] ugly, black dumbbells of all sizes lying in the slanted champagne racks." Each of the twenty rooms had a brass plate affixed on its door, on which was inscribed the name of a famous British writer. The Dickens room, for example, was the salon for medicine-ball work.

The perfect couple was a powerful draw, and No. 6 almost instantly filled with paying guests. It was only then that Macfadden got around to taking his bride, who'd kept their marriage secret for nearly half a year, to buy a proper set of wedding rings. Macfadden escorted Mary to what he considered to be the appropriate retail establishment at which to purchase a token of one's undying love: a pawnshop. He was incredulous when Mary left the store in tears. "Those rings would cost a pretty penny in London," he shouted after her. "They look like new. Nobody can tell the difference!"

One day in August, Macfadden told Mary the tale of a seven-year-old girl whose mother had come to his health home for treatment of a terminal disease, but too late to be saved. He'd made a deathbed promise to care for her daughter, he said. He tenderly asked Mary if she would help him raise her, and she readily agreed. The girl arrived in September. She was quiet and shy, with hair, her adoptive father proudly noted, the color of carrots. Bernarr named her Helen Macfadden.

To say that Macfadden raised Helen as if she were his own daughter would be to imply that she wasn't his biological offspring. Most of the people around him—Mary included, if one reads between the lines of her memoir—assumed she was the daughter he sired with Susie Wood, who remained at the Chicago Healthatorium.

Just after Christmas, 1913, Mary went into labor. Two mid-wives were summoned. Forty-nine excruciating (and doctorless) hours later, little Byrnece Macfadden entered the world, weighing five pounds. Her father, embarrassed by the puniness of his physical culture stock, added three pounds to the official birth notice.

On the fourth day of Byrnece's life, with a bitter cold whipping through the house, Macfadden took the baby into the bath-room, filled the basin with cold water, and proceeded to dunk the child "up to her nose and mouth" repeatedly. Mary grabbed her flailing baby away from its father and screamed, "Bern! For the love of Christ! Have you gone mad!"

Macfadden responded that Byrnece would be toughened up by the process, and demanded that it be done every morning. "When the baby is inured to the cold water, vitality will flow through its body," he insisted angrily. "The blood will race through its veins. . . . You're not going to interfere with my teachings and practices! You signed up!"

One month after giving birth, Mary was pregnant again. She would remain so for the better part of the next four years.

Five weeks into the life of the physical culture family, Mac-fadden returned home to No. 6 one night and announced that the family would be departing forthwith for Paris. The choice of des-tination was odd, to say the least. Macfadden considered the French to be a wine-swilling bunch of layabouts who poisoned their blood with fatty food and who were slowly committing "race suicide" by not bearing enough children. His sympathies were much more in line with those of the physically disciplined Germans and their martial peers from the Far East, the Japanese.

In the chill Paris winter, Mary came down with pneumonia, and though she was nursing Byrnece, the treatment her husband mandated combined fasting with hydropathy; her victuals were cut off, and her chest was hosed down daily with cold water. The family soon moved on to the small coastal town of Wimereux, on the straits of Dover, where Mary would look across the waves and pine for her native Britain. The Father of Physical Culture busied himself with traveling back and forth to London two or three times a week, dictating his memoirs to a secretary he'd imported from New York City and eating handfuls of fine quartz sand along the beach. Mary was shocked by the latter habit. Loyal *Physical Culture* readers would not have been, since Macfadden had already published two stories on the digestive magic of sand. One of them had been titled "Sand Cleans Glass Bottles—Why Not Bowels?"

In July, word arrived in Wimereux of Archduke Ferdinand's assassination and Austria's subsequent declaration of war against Serbia. The conversion of the French to physcultopathy would have to wait. The Macfaddens returned to Britain in September as the bloody Battle of the Marne was commencing a little over a hundred miles away. Macfadden had been receiving all-clear signals from the New York office, and his decision to return to the United States was probably hastened by a belief that the physically fit German troops would steamroll the corporeally less enlightened French. Macfadden quickly arranged to leave the Brighton home in the capable hands of his lieutenant, Stanley Lief, a graduate of the Chicago Healthatorium. Lief, himself a remarkable pioneer in the wellness arena, would go on to found the British College of Osteopathy as well as Champneys, a "top people's health farm" that thrives to this day.

By mid-October, the four Macfaddens were nestled in a second-class cabin aboard the RMS *Lusitania*, bound for New York City. The economizing was probably unnecessary, as Macfadden was wearing a money belt packed with $6,000, profits from the Brighton home. Word trickled down from the *Lusitania* crew that the scheduled five-day trip would take closer to nine; the ocean liner followed a zigzag route in order to avoid German

U-boats patrolling the Atlantic. (The steamship would make several such crossings until May 7 of the following year, when it was torpedoed and sunk.) When the pregnant Mary showed signs of seasickness, her husband cut off her solid food until the Statue of Liberty was in view.

Chapter 8

Lost Crusader

Hidden forces, sometimes marvellous and mysterious, lie within nearly every human soul. Develop, expand and bring out these latent powers. Make your body splendid, your mind supreme; for then you become your real self.

—Bernarr Macfadden, *Vitality Supreme*

America, and New York in particular, had turned upside down in the two years since Macfadden's departure in 1912. The uptight Victorian country he'd been chased from seemed to be loosening up, mutating, molting. An increasingly progressive Teddy Roosevelt had formed the Bull Moose Party to run for a return engagement as president, and he had been shot at for his efforts. A deep economic recession had brought on a rise in labor unrest. The deadly fourteen-month-long Colorado coal strike ended with National Guard troops firing on miners and setting their camps ablaze. Suffragettes paraded down Pennsylvania Avenue demanding the right to vote, and Illinois became the first state to allow women to cast ballots. In January 1914, Henry Ford helped to jump-start the American middle class by doubling his Detroit workers' wages to $5 a day. They'd earned it. Using efficiency expert Frederick Taylor's principles of scientific management, the time needed to assemble a Model T plummeted from twelve hours thirty minutes in 1913 to just ninety-six minutes one year later.

Manhattan seemed to be changing by the hour. The New York Armory Show that opened on February 17, 1913, introduced Manhattanites to works like Marcel Duchamp's *Nude Descending a Staircase* and ushered in the modern era in art. Two months later the sixty-story, 792-foot-tall Woolworth Building was completed on lower Broadway. Newly inaugurated President Woodrow Wilson christened the world's tallest building on April 24 by pressing a button in the White House that ignited eighty thousand lights.

Nowhere were the changes more pronounced than in the realm of sex, always slippery territory for Macfadden. Sigmund Freud had come to Massachusetts in 1909 to deliver a series of lectures (in German) at Clark College, and over the next few years, his ideas had trickled down to the general public. "Americans absorbed a version of Freudianism that presented the sexual impulse as an insistent force demanding expression," observed John D'Emilio and Estelle B. Freedman in their landmark history of sex in America, *Intimate Matters*. "The implications seemed clear: better to indulge this unruly desire than to risk the consequences of suppressing it." New York was the nexus of these changes. In May 1913, a copy of Paul Chabas's kitschy nude *September Morn* appeared in the window of a shop on Forty-sixth Street and went largely unnoticed until Anthony Comstock demanded its removal. Times were changing: The print remained on public view and became a sensation, drawing throngs of oglers and buyers. Eugène Brieux's syphilis drama *Damaged Goods* opened on Broadway. "A wave of sex hysteria seems to have invaded this country," observed the journal *Current Opinion* in August 1913. "Our former reticence on matters of sex is giving way to a frankness that would startle even Paris."

Macfadden sensed the changes immediately. "Revolution is in the air," he declared in a long *Physical Culture* interview he arranged to welcome himself home. "Reforms of all kind are vehemently demanding recognition. And notwithstanding the public inclination to characterize the leaders of reform movements as unreasoning fanatics, much of the truth that is preached by these investigators is being inculcated into the conventional life of to-day." The conversation sketched out Macfadden's immediate and long-range plans to help lead this societal peristalsis. His short-term agenda was to finish and promote a new book, *Manhood*, which would explain the vitality-sapping effects of masturbation (he said it caused acne as well as general nervousness, though not insanity, as some believed) and the withdrawal method of birth control (an abomination that he deemed "unnatural"), as well as revealing to the world how a man's thyroid gland could be massaged to return color to gray hair. Nor had Macfadden forgiven his old targets while abroad: "Constipation is a prevailing evil," he sternly told the reporter. Near the end of the interview he dropped a hint of things to come: "Political contests in which the opponents derive their support through advocating certain physical culture reforms will, I believe, become a reality in the not far distant future."

The Macfaddens stepped off the *Lusitania* and into a taxi. They were driven directly to Yonkers and the home of titular Physical Culture Publishing Company chief Charles Desgrey. Macfadden explained to Desgrey that he had reconsidered their deal and was going to restructure the business with Mary as his partner. This, regrettably, required him to renege on Desgrey's promised stake in the company. Mary was interested to learn that Desgrey was the brother-in-law of Susie Wood, her husband's former housemate at Physical Culture City.

The Macfaddens decamped to Long Island for a few weeks, then relocated to the International Healthatorium in Chicago. If Mary had not yet surmised the eminent position her husband held in the world of health, she surely got a clue when their party was met at the building's entrance by a muscled footman who greeted the returning founder like visiting royalty. Inside, the massive temple of fitness was like an Emerald City where Mac-fadden was the Wizard of Oz. The busy-bee staff consisted of "men and women in white uniforms, working together in the cause of the Master," Mary wrote. "They all looked like agile wrestlers with barrel chests, powerful hips, strong arms and low foreheads." The highest caste of employees wore Vandyke beards and spoke frankly about bodily functions at all times. "Their conversations often sounded like debates between obstetricians," Mary recalled.

In early 1915, the Macfadden family moved back east, renting a twenty-room summer home on Long Island. The spare rooms were quickly filled with boarders drawn through advertisements in *Physical Culture*. Macfadden commuted to the city in a 1903 Palmer-Singer car with a horn that played "In the Good Old Summertime." Beulah Macfadden arrived on April 19, 1915, weighing nine pounds. Macfadden was pleased by her size, though not as much as he was a year later, when his third daughter with Mary arrived. This child weighed almost *thirteen* pounds at birth. Her father's disappointment at having sired another girl was overcome by his pride in the results of his breeding techniques. Mary, long since resigned to the fact that her children's names would all begin with the letter "B," and perhaps sensing that her husband's ecstatic mood might provide her one chance to influence the choice, suggested "Brenda."

"Brenda!" Macfadden replied in disbelief. "Brenda is no name for this true physical culture baby! I'll call her BRAWNDA."

Delivering a trophy baby evidently gave Mary the right to an opinion, as the girl's name was eventually softened to "Braunda." Like Henry VIII, however, Macfadden had grown tired of daughters. And like King Henry, he would soon devise a plan to do something about it.

As *Physical Culture* approached its eighteenth birthday in 1916, the magazine was suffering from an identity crisis. Around the time Macfadden had departed for Europe, it lost its founder's voice. Instead of bellowing about great men, vitality, power, and efficiency, the magazine now lectured about how to solve personal problems.

To a large degree, *Physical Culture* had become a magazine for women. It redoubled its advocacy of physical equality with men in articles such as "Clean Houses or Clean Bodies?" The latter, of course, was Macfadden's priority. A May 1913 editorial titled "Getting a Square Deal for Woman" proclaimed: "Let us insist upon proper wages for the girl bread-winner . . . but a far more obvious and important duty is it to reform the abnormal and unnatural attitude of man toward what he regards as in every way the weaker sex." In the September 1913 issue, Charlotte Perkins Gilman—author of the classic protofeminist story "The Yellow Wallpaper"—wrote on "Motherhood and the Modern Woman," defending working mothers on the grounds that "the deepest life-process is not going to be disturbed by a mere change of occupation."

On top of this political feminism, however, editor John Brennan ladled women's lifestyle columns on planning menus and "housekeeping hints" of the sort that could be found in enormously popular women's magazines like *McCall's*. The mysterious and glamorous Madame Teru wrote in each issue on topics such as "Beautifying the Complexion" or suitable hairdos for ladies of a certain age. *Physical Culture* even employed an all-star cast of reformist thinkers to write on female-friendly topics. Mastication maven Horace Fletcher wrote a home economics column. Upton Sinclair, whose logorrhea evidently couldn't be cured even through fasting, continued to contribute serials such as "The Health of Little Algernon," about a sickly orphaned heir, and to write on topics such as "To Marry or Not to Marry?" British sexologist Havelock Ellis began writing regularly about contraception in 1915—the same year that his new friend, Planned Parenthood founder Margaret Sanger, returned from her own European exile. She'd fled the country after her arrest for violating the Comstock Act by distributing a newsletter that included information on venereal disease. In 1916, Sanger marked the historic dismissal of charges against her for distributing birth control by writing the pro-contraceptive *Physical Culture* essay "A Message to Mothers."

The issues of *Physical Culture* that came off printing presses in 1915 and early 1916 were particularly schizophrenic. A reader passing a newsstand could have grabbed a copy of a magazine featuring an apple-cheeked maiden on the cover promising to reveal the secret of radiant beauty and upon returning home would have found a long essay by George Bernard Shaw attacking the butchery of surgeons and idiocy of "microbe hunters." This period also saw a surge in photographs submitted by subscribers—usually shirtless and often pantsless as well—showing the world the fruits of a physical-culture lifestyle. To Macfadden's great regret, this gave *Physical Culture* appeal to a gay audience. But no matter how hard he protested against "painted, perfumed . . . mincing youths . . . ogling every man that passes" his love of the male body beautiful inevitably got the best of him. A full-page illustration that accompanied an episode of John Coryell's serial "Cyclone Adams: Athlete Detective" showed Cyclone standing before another man in the pose of Michelangelo's David, if David had added some crunches and lat pull-downs to his workout. Only a strategically placed towel hides Cyclone's manly plumbing. In the caption, the other fellow gushes: " 'Great Scott! What a body! I never saw such a balanced, harmonious development in my life. No wonder you can make sixty miles and come back almost as fresh as when you started! No wonder they call you Cyclone!"

Sometimes the magazine was merely carelessly edited, to unintentionally comic effect. An earnest 1916 article called for a statue of the pioneering Department of Agriculture chemist Dr. Harvey W. Wiley to be erected in Washington, D.C., in recognition of his "great fight for pure food." Adjacent to this rousing salute was a diagram of a toilet that would allow its user to inspect what he or she had left inside, for "it is just as essential to know what is being eliminated from the body as what is being put in." One editor's note begged forgiveness for misquoting a doctor on how a person might best take in some citrus after following a fasting regimen. "Instead of recommending an enema of fruit juice," the apology explained, "Dr. Levansin suggested the *drinking* of orange and lemon juices when breaking the fast."

Why did the magazine seem to be on a rudderless journey? One reason, obviously, was difficulty in keeping pace with the rapid changes in American society. Another was that Macfadden was occupied with other interests. He had arrived back in New York vibrating with new ideas, only some of them related to magazine publishing. His book *Vitality Supreme*, arguably the finest of all his published works (and the first in which he appeared on the cover in business attire), was published in 1915; in it he strove to convince the workers of America that they would soar to new levels of efficiency and success if they'd only adopt physcultopathic habits. Macfadden started the Vitalized Air Company, which installed air purification systems in buildings such as the Majestic Hotel at Fifty-fifth and Broadway. Influenced by the two-tiered red buses he'd seen in London, he attempted to market a double-decker subway car, which he believed would solve overcrowding as long as commuters didn't mind riding to work bent over. He patented an invention called the Peniscope. This device, which consisted of a glass tube and a vacuum pump, was intended to reinvigorate the hidden vitality centers of tired executives.

In October 1915, a one-page advertisement appeared in *Physical Culture* that read "Bernarr Macfadden in Motion Pictures: A New Physical Culture Course Illustrated in Unique Fashion." Macfadden, looking for inventive methods to preach the gospel of physical fitness—and with war looming, its growing importance to national security—had teamed up with the Universal Film Company (precursor to Universal Studios) to produce short films demonstrating proper exercise techniques. "All we ask of you is to watch the exercises carefully on the screen—read the instructions and explanations—and last but most important GET THE HABIT—and keep it up."

The first episode costarred and was directed by Allen J. Holubar, who also starred in the 1914 classic adaptation of *20,000 Leagues Under the Sea*. In it, a character representing Father Time walked onscreen and complained to Macfadden about "the deterioration of the race." Macfadden whipped out a copy of a book entitled *Building the Health of the Nation*, then proceeded to strike his repertoire of classical poses. He demonstrated some exercises, and ended the show with a don't-try-this-at-home "exercise feat." The films debuted nationwide in January 1916, with new episodes appearing every two weeks.

Macfadden knew that, as a mongrel magazine publishing advice on muscle building alongside beauty columns, *Physical Culture* lacked coherence. He appealed to his readers to help him sort things out. The Editor's Viewpoint column for March 1916 commenced with this surprising plea:

WANTED—A TITLE
Does PHYSICAL CULTURE as a title convey to those who know nothing of the magazine, the actual character of its contents? Can you suggest a title explaining in a word or two that PHYSICAL CULTURE is a magazine that stands not merely for physical, but for mental and moral culture as well? Ten dollars are offered for the best letter submitted to the editor on this subject.

The same issue featured a contest soliciting essays explaining "Why do you read Physical Culture?" A $25 prize was offered to this winner. "Mind you, we *don't* want commendation; we seek *frank criticism*," the unbylined but unmistakable prose pleaded. The contests were repeated the next month. The results were either unimpressive or Macfadden got cold feet about renaming his baby—in the end the name *Physical Culture* was selected as the winner.

Already perplexed subscribers must have been completely baffled when their copies of the July 1916 issue arrived. The bold, all-type cover was a new departure:

A MAGAZINE Working for the Health of the American People
THIS MEANS Working for Just Laws
For Just Administration
For Just Wages
Intelligent Government
EDITED BY JOHN BRISBEN WALKER

John Brisben Walker was one of the most famous magazine editors in America. A generation earlier, he had purchased *Cosmopolitan*, multiplied its circulation twentyfold, and transformed it into one of the great general-interest magazines of its day. (This was a half century before Helen Gurley Brown remade it yet again, this time as a sex manual for single women.) The contents of Walker's edition of *Physical Culture* were even more shocking than the wrapping. The prose was polished and the amateurish design had been cleaned up. Walker managed to keep the spirit of the original *Physical Culture* while taming its roster of editorial oddities into a respectable magazine. The August issue even included an article by Macfadden's hero Dudley Allen Sargent, the Newton of fitness.

And then with September, the changes vanished completely. A note from Walker atop the table of contents explained that he'd been promised a chance to promote his liberal social agenda while covering the world of physical culture, but that this promise had not been fulfilled. His parting with Macfadden was amicable, and while he didn't say so, inevitable.

Directly beneath Walker's farewell was an editorial titled YOUR OLD EDITOR AT THE HELM. The familiar bark had returned.

With Macfadden once again in charge, the *Physical Culture* editorial formula went back to basics. Much of the female focus was retained, but the softness of recent years was quickly expunged. Macfadden immediately announced a new contest to find the strongest woman in America, which would "deal a smashing blow to the pernicious theory that weakness is an essential and desirable aspect of womanhood." Margaret Sanger stopped tiptoeing around her pet subject with essays about motherhood and published "My Fight for Birth Control." Hollywood profiles were toned up to concentrate less on beauty and more on vigorous health. One of the year's longest features praised a new film starring "Drugless Douglas Fairbanks," in which the lead character helps a group of sanitarium patients recapture lost vitality through natural methods. In midyear, football coaching legend Walter Camp introduced the setting-up exercises that within a few years would launch the Roaring Twenties' first fitness craze, the "Daily Dozen." Even child-care advice was toughened up. In stories such as "A Kiddies' Gymnasium," Macfadden explained how to construct a portable, multistation exercise apparatus for building preschool muscles—just as he'd done for his own children. "If the time that is ordinarily used in dressing dolls and playing 'mother' to these inanimate objects were spent on the kiddies' gymnasium," he wrote, "there would be less predisposition to colds and other childish complaints."

America's entry into the Great War in April 1917 unleashed a torrent of editorials such as "A Patriotic Duty—Stop Overeating" and "Why Were the 'Doughboys' Denied the Benefit of Spinal Adjustment?" the latter of which demanded an explanation of GIs' lack of access to chiropractors. When he wasn't busy thumping his chest, Macfadden patted himself on the back. In his view, the Spanish Flu epidemic of 1918, which took 675,000 American lives, proved that those who poisoned their blood with white flour, alcohol, and medically approved drugs were susceptible to infectious diseases. According to Macfadden, virtually every sufferer who used physcultopathic treatments or sought care from alternative healers was cured. The December issue of *Physical Culture* included two self-congratulatory stories on the disease:

"Cause of the Epidemic—Denatured Foods" and "Spanish Influenza—Internal Cleanliness," written by Macfadden. "Influenza has brought death to thousands within the past few weeks," he explained with impatience and a hint of satisfaction. "Diseases of this sort are made possible only by one condition. An unclean alimentary canal."

As his fiftieth birthday approached in 1918, Bernarr Macfadden's beloved *Physical Culture* once again had an outward appearance of vigorous health, but its owner's finances were in criminally weak condition. Its year-end issue had boasted that 1917's "circulation gain has been greater than ever before." The locution "circulation gain" is commonly used by publishers to describe a situation in which readership has rebounded from a fall, and *Physical Culture* was actually selling the same number of issues as it had fifteen years before. Macfadden was still paying off the tens of thousands of dollars in legal debts that his aborted 1912 lecture was supposed to alleviate, as well as the printing bills he'd incurred while away in Europe. He had a wife and five children to support, his fourth daughter with Mary having been born on January 26, 1918. (Her name, Beverly, was taken from that of the road in Douglaston, Long Island, on which the family was living.) Yet he owned no home, had not a stick of furniture to his name. "Our living expenses were cut to the bone and the bone had been polished like a piece of ivory," Mary recalled.

But amid the tensions of 1918, a brilliant idea emerged for a new publication that would turn their lives upside down. And it would forever change American culture in the process.

Chapter 9

Nothing but the Truth

*There is nothing wrong with sex stories. No romance ever
existed that wasn't a sex story. No marriage is ever performed
that doesn't involve a sex story. It's only prudery to point a
forbidding finger at sex. You can't do anything about such an
attitude. You can only lift sex to its proper dignity.*

—BERNARR MACFADDEN, EXPLAINING HIS EDITORIAL PHILOSOPHY
TO JOURNALIST STANLEY WALKER

The year 1919 was the bridge between World War I and the
Jazz Age, a year when the Volstead Act instituting Prohibition and a bill securing a woman's right to vote passed in Congress within months of each other. It was the moment when, as
the great popular chronicler of the 1920s, Frederick Lewis Allen,
neatly put it, the trickling down of Freud's libidinous theories
mingled with the disillusionment of the Lost Generation and the
rising status of women to launch the first American sexual revolution. Chronicling and prodding this great unzipping was a new
form of literature, the confession magazine. That genre sprang
from Bernarr Macfadden's new publication, *True Story*.

The October 1916 *Physical Culture*—the first with the old
editor back at the helm—had included this plea to readers:

We Want Interesting Letters:
Liberal Remuneration for Letters Setting Forth Personal Experiences in Detail

Cash prizes were offered for the best stories. The ad even suggested some topics: "Confessions of a Woman Hater" and "Why
I Remained an Old Maid."

"We want the material in this magazine to be 'live,' " said the
ad. "We want it to come direct from those who have had actual
experience."

The solicitations continued throughout the following year: "Is
Your Marriage a Failure? Write Your Life Story for Us." This
time the magazine offered to pay $10 for each usable contribu-
tion. "Don't try to be literary," the invisible editor advised. "Write
your story frankly and simply, just as you would talk it. Truth is
stranger than fiction. It is usually far more interesting."

The true story of how *True Story* was invented comes in several versions. Macfadden told one writer that he'd gotten the idea
after visiting a convention of the publishers of pulp magazines, a
genre of sensational short stories printed on cheap wood-pulp
paper. He told another that he'd had a flash of inspiration while
strolling the beach at Wimereux, maybe between mouthfuls of
sand. One of his authorized biographers traced *True Story*'s gen-
esis back to late-night chats between Macfadden and John Coryell before Macfadden's involuntary European sojourn. Yet another
heard that Macfadden had scribbled the words TRUE STORY on
a piece of paper, held it out at arm's length, and imagined how
they would look peering out from a newsstand. In this version,
the visual aid sold him on his own genius.

Late in life, Mary Macfadden claimed that *True Story* had
been *her* idea. She'd sprung it on her husband, she said, in February 1918 while on a long walk a few weeks after Beverly's birth.
"Let's get out a magazine called *True Story*, written by its own
readers in the first person," she remembered telling her husband.
"This has never been done before. I believe it will have a wide

readership. It might lead to all sorts of publications in the same vein." In Mary's telling, which conveniently takes credit for all of her husband's future publishing success in just four sentences, Macfadden pondered her concept that evening while chewing a carrot at their crowded dinner table. Only when their guests started to praise the brilliance of her blueprint did he fully intuit its magnificence—and swear his guests to secrecy so that he could pursue it.

The question of who invented *True Story* may incite few barroom debates today, but the effect that this lightning bolt of a magazine had on American publishing—and American culture—was immense. The women's magazine market in 1919 was an exclusive, and wealthy, club dominated by a half dozen periodicals known as the Big Six. (Two of the six grannies, *Good Housekeeping* and *Ladies Home Journal*, have endured into the twenty-first century.) These were *serious* magazines; it was no coincidence that Theodore Dreiser, author of *Sister Carrie* and *An American Tragedy*, was the editor of the *Delineator*, a Big Six title in the early decades of the century. Each magazine offered the same general contents: home economics tips, family advice, literary fiction, and long pieces of investigative journalism. All were aimed at a decidedly upscale audience.

True Story was unlike any of them, nor was it like any publication then in existence: a magazine devoted entirely to nonliterary, apparently factual stories told in the first person. The initial issues appealed to both men and women. One early edition included not only the confessions of a "Washerwoman millionaire" and "the romance of a girl athlete," but also the story of "the man who outran a locomotive" and a science-fiction prophecy about Germany's future. The best tales, however, skewed toward the interests of America's exploding class of working girls: fallen women, lost loves, weekends abandoned to bootleg gin, and other "missteps." Most stories oozed with sex but also contained a moral lesson. *True Story* was, in a way, the nonfiction descendant of "Growing to Manhood," the tale that had caused Macfadden so much legal trouble a decade earlier. It would be the making of Macfadden's fortune and his final triumph over prudery.

To execute his vision, Macfadden turned to former *Physical Culture* editor John Brennan. Brennan was to have no say on the choice of stories, which would instead be selected by a staff of untrained readers "consisting of stenographers, dancing teachers, even wrestlers," who would grade every story submitted. Macfadden equated crudity with verisimilitude. Big Six magazines had run bylines such as Willa Cather, Rudyard Kipling, and Mark Twain. In *True Story*, professionalism was aggressively weeded out. "You can fix the grammar, you can punctuate and make the verbs agree with the subjects—but that's about all," Macfadden told Brennan. So what, Brennan wanted to know, would he be doing if not any actual editing?

"Get the book out—on time," Macfadden told him. "That's all."

The cover of that first issue featured a couple sharing a smoldering glance over the line AND THEIR LOVE TURNED TO HATRED. "We pay $1,000 for your life romance," promised a line at the bottom. The story mix would soon include "chatty bits on stage and screen personalities" and long romantic serials penned by stalwarts such as John Coryell and Milo Hastings. The few pages of advertisements were populated mostly by musclemen and the makers of breast-enhancement products.

A list of titles from that era gives some of the flavor of *True Story*:

Trapped by Destiny
Did I Do Right?
I Wanted Love and So—
Things Wives Tell
Ignorant of Life

As for the "truth" of the tales related in *True Story*, Macfadden's standards were less than exacting. Miraculous coincidences and statistically improbable reunions that would make soap opera producers roll their eyes were fairly common. But as far as Macfadden was concerned, if the writer said the story was true, it was true.

True Story's business genius was Orr J. Elder, one of Macfadden's first hires at *Physical Culture*. He, too, helped to turn the Big Six model on its head. Elder charged a hefty twenty cents per copy at a time when the *Ladies' Home Journal* cost fifteen cents and most magazines could be purchased for even less. *True Story* was the sort of impulse purchase that sold almost exclusively through newsstands rather than via discounted subscriptions— the magazine sold a phenomenal 96 percent of its copies on newsstands—sending any advertising revenue straight into Macfadden's pocket.

True Story's sales started modestly with about fifty thousand copies of the May 1919 issue. It quickly became clear, however, that Macfadden had tapped into the vein that every media baron dreams of reaching: an entirely undiscovered audience.

The November 1919 issue of *Physical Culture* celebrated two milestones. A two-page spread of photographs of the nearly nude editor displaying his impressive physique ran under the headline "Bernarr Macfadden and His Athletic Fifty-First Year." The montage was meant to "show very clearly that the founder of this magazine is now in just as perfect condition as he was twenty years ago, or perhaps even a little improved." A few pages later appeared an advertisement inviting readers to buy stock in "Physical Culture Magazine and its allied interests." Twenty thousand shares were offered, and quickly sold, at five dollars per.

Thanks to *True Story*, the Physical Culture Publishing Company more than tripled its annual revenues between 1918 and 1920, breaking a million dollars for the first time. With money suddenly flowing in, Macfadden finally purchased a home, a ten-room house in Nyack, New York, up the Hudson River from New York City. He ordered several sleeping alcoves added to the austere exterior, which Mary said gave the home, which she dubbed the Palace of Porches, the look of a Swiss chalet. "As it would be with every Macfadden place," she remembered, "the grounds were soon transformed into a fresh air gymnasium, with swimming pool, tennis courts, miniature-golf course, and booby-trapped with chinning bars, swings, trapezes, punching bag props, dumbbell parapets, and a dell for carrot chewing guests." The area was rural enough that the grounds were infested with copperheads. Macfadden often walked six hours and twenty-five miles to his Manhattan office. Tutors had to be employed for the children because they had not been vaccinated.

Mary had been granted two years of leave from baby making, with the help of family-planning advice from Margaret Sanger, to whom Macfadden had been quietly sending money for several years. During her furlough, Mary was a regular presence at the *True Story* offices, reading the tales of woe that had passed the screening committee. As the next-to-last person to weigh in on each manuscript, she proudly scribbled her initials on every story before handing it off to her husband. Often the Macfaddens would read stories aloud to each other, she in her English Midlands accent, he in a voice like W. C. Fields with a sinus infection.

Near the end of 1920, Macfadden invited Dr. David H. Reeder, an Indiana physician, to visit him in Nyack. That Macfadden would host a physician was strange, but Reeder's specialty was dear to Macfadden's heart. Reeder claimed to have discovered the secret to predetermining the sex of a child. His theory was this: "If conception takes place within three days after the menstrual flow has ceased, the offspring will be a girl. On the eighth day or any time thereafter up to the twentieth day, the offspring will always be a boy. I have never known this law to fail."

In the spring of 1921, Mary's office duties were reduced in

preparation for the couple's new breeding experiment. During her time at home in Nyack, she leafed through some back issues of *Physical Culture,* where she read for the first time Bernarr's earlier theories on sex determination, published during her first pregnancy. One of his ideas was that boys were more likely to be born to starving mothers, which explained his eagerness to cut off her victuals during prior pregnancies. This time, perhaps because he felt he'd stacked the deck in favor of a boy, Mary was allowed to gorge herself on roast beef.

As Mary's late December due date approached, Macfadden packed up the family to spend the holidays at an apartment they kept in the city, on Riverside Drive. Christmas was typically a spartan affair in the Macfadden home. Perhaps because of his unsentimental childhood, Macfadden was generally unenthused about toy giving. He felt that Santa was a bad role model, with his fat belly and tippler's nose. This year, however, between the anticipated arrival of a son and the runaway success of *True Story*—circulation was close to half a million copies per month and rising quickly—he was in a celebratory mood. He spent the night of December 28 in midtown hosting his company's holiday party, a decidedly non–physical culture event. As a physculto-path, Macfadden may have opposed the consumption of alcohol, but as a libertarian he was even more opposed to Prohibition. "When it came to smoking or drinking," an employee once explained, "his philosophy [was] that if people, even close friends, were satisfied with shorter lives their way than longer ones his way, that was solely their business."

When he returned home at one a.m., he found Mary about to give birth. After failing to secure a midwife, he delivered the baby himself.

"*It's a boy!*" he shouted in elation. "My theory is correct!"

Mary had delivered another thirteen-pound prize. The baby was named Byron, and an announcement of his birth was rushed into *Physical Culture.* After much harassment from a nurse who arrived after the birth, Macfadden permitted a doctor to enter

the apartment and give Mary fifteen stitches. Her pleas for anes-
thetic were denied.

Macfadden's businesses were expanding so rapidly that
the Fortieth Street headquarters were overrun. An office visitor
recalled that while the Physical Culture Publishing Company occupied a full floor of a large building, "Many of the offices that
had once been eleven feet high were now only five and a half feet
high, with a ladder running up the side into an upper compart-
ment where some editorial clerk labored in a cramped position all
day long." The owner of the company was shoved into a tiny
room "barely large enough for a desk and two chairs." Adjacent
to Macfadden's office, however, space was reserved for a private
gym. Macfadden frequently halted editorial conferences to grap-
ple with wrestler George Bothner, the former world lightweight
champion. "Editors waited for an hour or two while a series of
hard thumps resounded from the mat," Mary recalled. "When it
was over, Bernarr pranced back to his desk, often bared to the
waist, full of quick editorial decisions."

After twenty-plus years of publishing *Physical Culture*, Macfadden had become the national fitness spokesman he'd always
dreamed of, nearly as influential as John Harvey Kellogg and as
popular as Horace Fletcher. He made his first appearance in
Who's Who, where he listed his primary occupation as "founder
of physcultopathy." His daughters, known as the Macfadden-
ettes, were rousted from bed weekdays at four a.m. to make
appearances on New York's WOR radio, performing fitness routines for the growing audience of that fledgling medium. His
health demonstrations were no longer ignored by the mainstream
press; he would shortly embark upon a thirty-day fast that would
receive positive coverage in the *New York Times*, and somewhat
less complimentary coverage in the cheeky new weekly *Time*,
whose young editors were obsessed with the man they nicknamed
"Body Love" Macfadden.

Through *Physical Culture*, Macfadden had become the pied piper for a growing number of cultish fitness practitioners. Alternative healers and renegade MDs found a welcoming home in the magazine's columns. Natural-food wholesalers and hucksters promising to fix everything from baldness to bowed legs to bowel problems populated its advertising sections. Month after month, the bottomless gaze of Swami Yogananda, purveyor of meditation and future author of *Autobiography of a Yogi*, stared out at readers who perused the ads in the back of *Physical Culture*.

Macfadden was in particular a hero to bodybuilders. The sculpting of one's anatomy for purely aesthetic reasons, thanks largely to Hollywood, had progressed a long way from Macfadden's first physique demonstration in 1905. A February 1923 story titled "Red Blood and Plenty of Sand" introduced readers to the impressive corpus of the young Italian actor Rodolph Valentino. (He hadn't fully completed his transition from Rodolfo to Rudolph.) To make absolutely clear that the muscular legs on display in *Physical Culture*'s pages were not "effects contrived by trick photography," the article's author explained that "both Mr. Macfadden and myself sounded the firmness of Valentino's muscles with our own hands."

Largely due to Macfadden's cheerleading, the pioneering influence of Eugen Sandow and other Victorian musclemen had trickled down to a new generation. Some of them were content to send in near-naked photos of themselves flexing, to be printed in the monthly photo gallery called The Body Beautiful. Some sought an audience with the master. Writers and editors were accosted in the office halls by strongmen who demanded that staffers feel their bulging muscles, watch them bang out over-the-transom pull-ups, or perform feats of strength. No telephone book was safe from being torn in half.

Though an all-star team of early American bodybuilding could be assembled from the advertisements in *Physical Culture*— bare-chested behemoths such as Lionel Strongfort and Earle Liederman taunted the fragile egos of readers in every issue—none was so well known as a young man from Brooklyn who won *Physical Culture*'s "World's Most Handsome Man" contest in 1921. Charles Atlas was an Italian immigrant from Calabria, born Angelo Siciliano, who'd come to America at age eleven. By his late teens, he was working as a Coney Island strongman and artist's model. He'd picked up his first copy of *Physical Culture* in 1909 and made his first appearance in the magazine in June 1915, posing to illustrate a story on the proper dimensions of a real man. Macfadden was floored by Atlas's development in the photographs he'd submitted to the "Handsome Man" contest. He asked for a meeting.

Atlas arrived at the Fortieth Street offices, accepted a glass of carrot juice from Macfadden, stripped to a leopard-skin loin-cloth, and proceeded to strike a series of poses. Macfadden instantly named him the winner of the contest, handed him a check for $1,000, and declared that he was "the living realization of my lifelong battle for the body beautiful." Atlas's life story appeared in the November and December 1921 issues of *Physical Culture* under the title "Building the Physique of a Greek God."

In October of the next year, Atlas entered Macfadden's "World's Most Perfectly Developed Man" competition, held as part of the National Physical Culture Exhibition at Madison Square Garden. More than 750 musclemen gathered to compete for the title. Macfadden quickly declared Atlas the winner, handed over another $1,000 check, and canceled any further contests. "What's the use of holding them?" he asked. "Atlas will win every time."

Shortly after this second victory, Macfadden assigned one of his employees, a marketing man and bodybuilder named Frederick W. Tilney, to direct a movie starring Atlas. While chauffeuring Atlas between studios, Tilney suggested that they team up on a mail-order course. The Charles Atlas legend factory claimed that the strongman had hit on his system of "dynamic tension," in which one builds strength by pitting one muscle against the other, while watching a lion stretching at the zoo. Perhaps it is merely a remarkable coincidence that the first several years of advertisements for Atlas's mail-order strength course offered a sixty-four-page booklet titled *Secrets of Muscular Power and Beauty*, and that the ads began appearing almost immediately in *Physical Culture*. The title is virtually identical to that of Macfadden's 1906 treatise *Muscular Power and Beauty*, in which he also writes about growing strong through isometric exercise—that is, without heavy weights. Atlas's brochure also contains a great deal of information about building "radiant health," avoiding doctors, and consuming enormous quantities of milk. Prominent references were made to the criminal nature of weakness. In return for Macfadden's help jump-starting his career, Atlas advertised in *Physical Culture* for years to come, first as "Prof. Chas. Atlas," and much later as the savior of 97-pound weaklings.

A new Macfadden magazine appeared in 1921, dedicated to pumping up America's underdeveloped cerebrums. It was called *Brain Power*. Inside the debut issue were four stories penned by a man whose name was unfamiliar to Macfadden. The owner of the company, a man who, according to an associate, "thought editors took bribes, split checks with authors, and thus would rather buy trash from their confederates than good material from writers who refused to check up," was intrigued as he made out four simultaneous payments to this stranger. He invited the writer up to take a noontime stroll near his office.

Charles Fulton Oursler was the managing editor of the Broadway news sheet *Music Trades*, though that job title did not begin to scratch the surface of his abilities. At age twenty-eight, he was a self-taught polymath who seemingly had little in common with the health guru from Missouri. Tall and slim, with slicked-back hair and a large, elegant nose, Oursler drank moderately and smoked heavily. He was a serious amateur magician who hobnobbed with the likes of Houdini and used his wide knowledge of the occult arts to expose fraudulent mediums and clairvoyants.

Virtually everyone who worked with him remarked on what one subordinate called his "penetrating eyes," which constantly scanned the room, not missing a detail. Fiercely ambitious and driven, he had left his hometown of Baltimore a few years earlier, wife and children in tow, in order to pursue a literary career in Manhattan. A fondness for milk was the sole interest he shared with Macfadden.

Their first ambulatory meeting took place on the paths of Bryant Park, behind the New York Public Library. The two strolled and talked for hours about everything from politics to jazz to Freud. Macfadden was evidently impressed; by year's end he offered Oursler a job. Oursler declined. "Then Macfadden made me another proposition," Oursler recalled. "If I would spend an hour in the office with him every day at noontime he would pay me one hundred dollars a week." Within six months, Macfadden had worn him down, and Oursler came to work for him in June of 1922. He would from that day forward be known simply as "Fulton Oursler," though Macfadden called him "Charley."

There are several types of magazine editors, and a successful publishing company needs them all. There are editors who toss out ideas like confetti, or who know lots of writers and people of interest. There are editors who can dismantle and reassemble a troublesome piece of prose as a good car mechanic does a faulty engine, and send it off into the world purring. There are editors whose strength is writing headlines and photo captions so intriguing that they dare a reader to turn the page; editors who specialize in managing the odd, flammable mix of creative egos that are crammed into a magazine office; and magazine editors who are themselves great writers. Solid careers have been built on any one of these skills. Lucky editors possess two or three of them. Fulton Oursler had them all, plus a heightened sense of office politics and discretion. He once told his eldest son, "I guide my life by five words: dare, do and keep silent."

The staff was somewhat less excited than their boss about the arrival of their new colleague. Oursler's job description was vague, and he spent his first two weeks sitting alone in his new office, tapping out pieces for *True Story* and waiting for his telephone to ring. Not one employee stopped in to greet him. On his third Monday, he arrived to find two envelopes waiting for him on his desk, addressed in Macfadden's oversize, confrontational handwriting.

> Dear Mr. Oursler,
> I am going away on a trip and am leaving you in complete editorial charge of my properties. You know what I want. Give them their orders. I will back you up. You are herewith appointed supervising editor of Macfadden Publications.

The second envelope held an announcement of his promotion, signed by all Macfadden Publications editors. Had Macfadden hired Oursler to serve as his editorial Cardinal Richelieu, as Mary later claimed? Macfadden certainly had big plans for his business. On June 13, 1922, the company had announced that it was changing its name from the Physical Culture Publishing Company to the more stately Macfadden Publications. A few weeks later, the *New York Times* reported that Macfadden had signed a lease on a six-story building at Broadway and Sixty-fourth Street, on Manhattan's Upper West Side. The building was a reflection of Macfadden's own Janus-faced personality, sandwiched between verdant Central Park to the east and the sooty, rumbling Ninth Avenue elevated train to the west. The owner's grand new office had doors on either side, one leading to his executive secretary's office and the other to Oursler's. The new supervising editor could usually be found behind a cloud of smoke and took delight in lighting his cigarettes from matches struck on the No Smoking sign.

True Story was far too successful to keep what one wag dubbed the "I'm Ruined! I'm Ruined!" publishing field to itself. Fawcett Publications' *True Confessions*, which debuted in October of 1922, was a transparent copy and an instant success.

Around this time, Macfadden secured a million-dollar loan to expand his *True Story* holdings. In short order, Macfadden Publications spun off *True Romances*, *True Experiences*, *True Marriage*, *True Love,* and, in a slightly different vein, *True Detective Mysteries*. The name of each was engraved on a brass plaque on a pillar in the lobby of the new headquarters. The company's other new magazines with a physical-culture slant included *Dance Lovers* and *Muscle Builder*, a magazine devoted entirely to the pursuit of strength and size. Soon a row of pennants, each bearing the name of a Macfadden magazine, flew atop the building.

Success may have made Macfadden more vain, but it did not affect his personal appearance. He still frequently walked to work barefoot in a rumpled suit, wearing a fedora punctured with dozens of holes, which he felt allowed his brain to breathe but which his daughters called his "gang-war" hat. Legend has it that Macfadden arrived for work one day and began to castigate a new porter for his poor sweeping. The porter, who'd never met the company founder before, attempted to eject the shabby, loud-mouthed intruder whose hair stood on end as if shocked. The two ended up wrestling. Macfadden had just vanquished his opponent when Rudolph Valentino, arriving for a photo shoot, walked into the lobby and said, "Good morning, Mr. Macfadden. Taking a little exercise?"

Flags rose and fell on the Macfadden Publications Building roof in a frenzied semaphore. *Brain Power,* marketed to those seeking parole from intellectual weakness, held limited appeal to advertisers and stumbled along for a couple years before vanishing. *Beautiful Womanhood,* a pet project of Mary's aimed at an audience that saw itself as challenged in pulchritude, met a similar fate. The short-lived *Midnight Mystery Stories*, a publication that told chilling and salacious tales of life after dark in the Naked City, debuted in August 1922.

Midnight disappeared suddenly in early 1923; Mary claimed years later that her husband had shut it down immediately when John S. Sumner, Anthony Comstock's successor as head of the vice squad, arrived at Macfadden Publications with summonses for everyone associated with the magazine. It is probably no coincidence that lawmakers were proposing legislation that called for automatic life imprisonment for anyone with four felony convictions. Macfadden had two.

In January of 1923, Macfadden purchased the sagging *Metropolitan*, which just a few years before had been publishing the work of former president Theodore Roosevelt and of John Reed, author of *Ten Days That Shook the World*. Oursler made his aspirations for the magazine clear when he paid $3,000 for an unpublished Theodore Dreiser novel, *The Genius*, as the showpiece of the first issue. Macfadden insisted that while Dreiser's work was nice, the lead story should be a *True Story* submission, "The Life Story of an Artist's Model." Oursler was well aware of the article's modest literary merits—he'd written it. Both stories ran in the issue. *Metropolitan* carried on for a year before dying.

On December 4, 1922, Macfadden returned home from work to his Riverside Drive apartment. He was greeted by the sounds of happy children running about and a player piano chirping out the notes to "Babes in Toyland." Eleven-month-old Byron, known within the family as "Billy," was seated on his mother's lap. Suddenly, Billy tensed up, threw his head back, and began to contort his body as if overcome by a seizure. Bernarr demanded that the infant be stripped and dunked in a steaming sitz bath. Mary recalled that the water's temperature was so hot that she couldn't keep her hand submerged. Though it is impossible to know the cause of Billy's fit, many common childhood seizures are now known to be brought on by fever, so a hot bath as treatment was probably ill-advised at best. The baby's spasms continued. Mary snatched him out of the water and screamed, "Bern, for the love of Christ, call a doctor!" Billy died in her arms.

Macfadden prescribed a physical-culture remedy for the family's grief: exercise. "I'll walk with you for a week until you all get a good, healthy, tired feeling," he told Mary and the girls. The entire family—except for Helen, who was attending a girls' farm school—decamped to Atlantic City, where day after day the six of them trooped up and down the snowy boardwalk, with Macfadden setting the pace.

Macfadden had publicly bragged for almost a year about the breeding triumph that was Billy. Always a little paranoid, he now expected a backlash, perhaps violent, as word of the boy's death leaked out. Upon returning to Riverside Drive, the family boarded a train for Athol, Massachusetts, the small town near the New Hampshire border where Helen went to school. They all passed a dreary Christmas at the schoolmaster's home. With the girls out of harm's way, Macfadden convinced the distraught Mary that the only way to keep her overwhelming sadness at bay was to keep walking—all the way back to Manhattan, through the snow, two hundred miles away. The two packed light luggage and departed into the subzero weather. Macfadden phoned in to the New York office at stops along the way, one time receiving the news that the first issue of *True Romances* had sold out. At an inn in Greenwich, Connecticut, he persuaded Mary to try for another son, for the sake of her emotional stability and his reputation. Amazingly, she acceded to her husband's wishes that night. The couple then caught a train back to Manhattan.

At odd moments along their trek, Mary had noticed her husband scribbling words on scraps of paper. Upon his return to the office, Macfadden showed Oursler an editorial he had written about his son's death that he wanted published in *Physical Culture.*

Billy was often over-fed . . . I protested on numerous occasions . . . but my protest was not vigorous enough. Anyway, I believed the boy was so strong that he would overcome mistakes of that nature. And it is so hard to combat the tendencies of mother love. . . . I also somewhat blame myself for neglecting his exercises.

As diplomatically as he could, Oursler suggested that perhaps a national magazine was not the place to blame a baby's mother for his death. Macfadden ignored his advice, on the grounds that his nation of physcultopathists deserved an explanation. Without informing Mary, he ran the commentary in the next issue. Thousands wrote in to congratulate their hero for sticking to his guns. Macfadden was obviously proud of the editorial—enough so that he reprinted it, in full, in two of his three authorized biographies. For Mary, the shock of reading it opened a schism in their marriage. "A deadly curtain . . . was coming down on us," she later wrote, "to put me and the children on one side of it, and Bernarr on the other."

Nineteen twenty-three was a banner year for Macfadden Publications. Profits doubled, largely behind the phenomenal growth of *True Story*, which would reach a circulation of eight hundred thousand by the end of 1924. *True Romances* was selling strongly. Macfadden's role as a health crusader was buttressed by his success as a businessman—the company was now regularly announcing frequent generous dividends. When the U.S. Congress and the New York State legislature, prodded by John S. Sumner, began talking about passing bills that would have permitted government-sponsored vetting of magazines as well as books, Macfadden stood with literary lions such as George Palmer Putnam of G. P. Putnam's Sons and Alfred Harcourt of Harcourt, Brace and Co. in attacking the censorship of printed material. The *Times* quoted at length a telegram Macfadden sent to Senator Salvatore A. Cotillo, chairman of the Senate Judiciary Committee. "If your bill becomes a law every publisher who deals with sex in fiction, instruction or in any form in New York State will be in hourly fear of arrest regardless of how exalted his motives may be. The passage of your bill would compel me to immediately consider moving to another state." At the rate Macfadden Publications was growing, such threats were now taken seriously.

The family's grief was eased—and Macfadden's sex-determination theory vindicated—in October, when Mary gave birth to yet another thirteen-pound boy. His father named him Berwyn, after a town in western New York. The year closed out with a silver-anniversary gala, held on December 19 at the Hotel Astor, to celebrate twenty-five years in publishing. "Dance numbers were interspersed with acts from various Broadway attractions," said a society reporter's recap of the evening. "There also were physical culture and fencing exhibitions." The odd mix of entertainments may have reflected the unusual creature the evening's honoree was planning to release into the world in 1924.

Chapter 10

The Porno-Graphic

Macfadden announced the forthcoming Graphic *in page
advertisements in the New York papers as the most unique
daily that would ever be seen since Johannes Gutenberg did his
first printing. This claim was not exaggerated.*

—Emile Gauvreau, *My Last Million Readers*

In 1919, a British invader landed on the island of Manhattan:
the tabloid newspaper. Five years later, it had become a permanent fixture in America's largest city. The tab was the creation
of Alfred Harmsworth (later Lord Northcliffe), whose lowbrow
London *Daily Mail* had earned him a fortune at the turn of the
century and made him one of the most influential men in England. At Northcliffe's urging, his friend Joseph Patterson, head
of the family that published the *Chicago Tribune*, launched the
Illustrated Daily News on June 26, 1919. To New York newspaper buyers accustomed to broadsheets such as the *Sun* and *Times*,
its appearance must have been shocking: half the size of those
dailies, with a front page devoid of a single column of sober type.
Instead, one large-type headline blared over a sole arresting photograph, or a handful of photographs. Unlike the broadsheets,
the tabloid did not need to be unfurled like a horse blanket and

could be flipped through easily on the subway. As the historian and publisher Simon Michael Bessie noted of the nation, and doubly so for the de facto capital of the Jazz Age, "America was in a hurry—for big money; for greater speed; for more fun; fun in cars, fun in speakeasies, fun in jazz, fun in movies, fun in newspapers."

Within two years the *Daily News*, as it quickly came to be called, was the second-largest-selling paper in America's number-one media market, moving four hundred thousand copies a day. The number was growing rapidly.

One morning early in 1924, Macfadden summoned Oursler into his office. "For many a long day I have wanted to publish a newspaper in New York," he told his editor. "Now I've got the money to do it. *True Story* for the last year has been earning net, after taxes, about ten thousand a day. I think we can get it ready. Please find me an editor, and start building a staff."

With Macfadden's instincts for mass audiences, there was no question that the *Truth*, as he planned to call it, would be a tabloid. He had two dictates for the *Truth*: Promote physical culture—"That's why I'm starting it," he said—and "editorialize" the news. "Don't stop with the bare skeleton of the facts," he explained to one job applicant. "Point out the moral—the social lesson. A man's found dead, from an alcoholic spree. Don't stop with saying it—drive it in! Influenza epidemic—play up the fact that the bodies of the victims are susceptible to the disease, because of lack of bodily care—overeating, and all that. A girl is picked up by a stranger for an automobile ride, is assaulted and beaten. Play up the social moral—did the mother teach her properly, as to avoiding such contact with strange young men?"

In his public announcement, Macfadden's own précis of the *Truth*'s formula was more succinct. "This paper will be devoted largely to self-help," he said.

Macfadden wanted Oursler to oversee the paper, but he had plenty on his plate at that moment: He was editing *Metropolitan* with a tiny staff, supervising the other five Macfadden magazines, and churning out more slick magazine prose than most full-time writers could. In his spare time—late nights and Sundays—he managed to squeeze in writing his first novel, *Behold This Dreamer*, which was published in 1924 to good reviews. Macfadden wanted his paper up and running quickly, though. William Randolph Hearst, America's king of sensational newspaper journalism, was rolling out New York's second tabloid, to be called the *Daily Mirror*, in June of 1924. He already owned two newspapers in the Gotham market, the *New York American* and the *New York Evening Journal*. Hearst's *Mirror* editor was boasting that the new tabloid would consist of 90 percent entertainment and 10 percent news.

So it seemed fortuitous that shortly thereafter, a small man with a mangled leg limped into the Macfadden Publications offices. His name was Emile Gauvreau, and he had been until recently the editor of the *Hartford Courant*, one of the oldest and stodgiest papers in the nation. Gauvreau was twenty-nine years old, of French-Canadian stock, and wore a three-inch lift in one shoe to compensate for an injury he'd suffered as a child. (Depending on where and when Gauvreau was telling the story, the cause of the maiming at age five had been either an accident involving a cannon-like firecracker or an actual cannon, which enemies of his father had fired into the Gauvreau home on the Fourth of July.) He had "a kindly feeling" for *Physical Culture* because exercises he'd taken from the magazine had allowed him to regain the use of his lame leg. Gauvreau, who'd contributed some lucrative stories to *True Story*, was in town for a meeting at the *New York Times*, and because his hotel was across the street and he had the afternoon free, he popped in to see who was around. He was immediately escorted in to see Oursler, who in turn took him to see Macfadden. By sundown, to his own amazement, he had signed on as editor of the *Truth*.

On the surface, the match appeared to be an excellent one for all three parties. Macfadden was familiar with Gauvreau's work investigating medical diploma mills in Connecticut, which had cost him his job. In Macfadden's view, anyone who hated doctors was worth hiring. Oursler was attracted to Gauvreau's experience running a conservative paper; he would presumably act as "a checkrein of good taste" on his boss's wilder inclinations.

Had Oursler been in less of a hurry, he might have delved deeper into his new employee's body of work at the *Courant*. Gauvreau was a newspaper obsessive who'd worked his way up from legislative reporter to managing editor in just three years. Oursler assumed that because Gauvreau had run one of the nation's most boring daily broadsheets, he must be a serious young editor with ambitions to put out a newspaper that reflected his sober values. Actually, Gauvreau's buttoned-up bosses had fired him because of his unwelcome efforts to juice up the *Courant*, as in a headline over a story about a surgeon, from the diploma-mill series: THREE CANS OF ETHER SNUFFED OUT LIFE AS SUTCLIFFE HACKED AWAY.

Gauvreau realized that he was a long way from Hartford on his second day of work on Sixty-fourth Street, when the office windows were thrown open and a well-toned gentleman in a loincloth leaped atop a desk at the front of the cavernous newsroom to lead the Macfadden Publications staff in calisthenics. Upon retiring to the fire escape, Gauvreau was informed by other employees who sought refuge there that the exercises were optional. He received a second surprise when he paid a visit to his future work site, the old *New York Evening Mail* building near City Hall. When he arrived, Gauvreau found a run-down red-brick structure, whose contents consisted of some rusted-out machinery and a flag-draped coffin, a memento from the wake held at the *Mail*'s demise.

But the dilapidated plant was whipped into shape quickly, for Macfadden had decided that September 15, 1924, would be his paper's launch date. The afternoon paper-to-be's name was changed, on the advice of a business manager, from the vaguely evangelical the *Truth* to the more advertiser-friendly *New York Evening Graphic*.

Macfadden's plan, which must have made more sense when he announced it than it did later, was for his newspaper to adopt the successful first-person formula of *True Story*. "As applied to the *Graphic*," Gauvreau explained, "the account of a man who killed his wife was not to be written in the third person from a police report. The prisoner was to be interviewed and his confession printed under his own signature." The headline over such a story, Gauvreau suggested, might have been:

I MURDERED MY WIFE BECAUSE SHE COOKED FISHBALLS FOR DINNER
I Told Her I Would Never Eat Them Again
but She Defied Me to the End
By Jonathan Peters

On September 15, after just two rehearsal print runs, the first issue of the *Graphic* appeared. The newspaper's motto running atop the front page read, "Nothing but the truth."

Macfadden's inaugural editorial declared: "We want [this newspaper] to throb with those life forces that fill life with joyous delight. *We want to show our readers how to live 100 per cent.*"

The contents themselves bore little or no resemblance to the news carried by the competition in the same day's editions. Where the *New York Times* featured page-one coverage of the presidential campaign and France's lowering the interest rate on Germany's war debt, the *Graphic* splashed the headline HERO SAVES PALS IN BLAST across its front page. The story, told by a foreman who'd saved three coworkers from an explosion at a dig site, "was considered routine news in other [newspaper] offices," opined the industry's custodial weekly, *Editor and Publisher*. Other headlines from that premiere edition may have set a record for pronoun use:

FRIENDS DRAGGED ME IN THE GUTTER
WE FACED DEATH TOGETHER IN THE FLAMES
I AM THE MOTHER OF MY SISTER'S SON

If the *Graphic*'s news coverage was unusual, the rest of its twenty-four pages bore little resemblance to any publication that had ever appeared anywhere. Bernarr Macfadden oversaw the physical culture page, Keeping Fit. Mary contributed a column for women. A *True Story* confessional ran each day. Films and plays were evaluated by two reviewers, the paper's professional critic and a man on the street, whose opinions ran side-by-side. Nearly every page trumpeted a contest of some sort. In one, readers were told to be on the lookout for Miss Courtesy, the paper's subway-roving correspondent, who awarded prizes to chivalrous readers willing to give up their seat. (To eliminate any danger of mistaken identity, the *Graphic* ran a full-page photo of Miss Courtesy in a bathing suit.) Another contest sought to find New York's most beautiful secretary. An ad for a competition dear to Macfadden's heart announced:

GRAPHIC SEEKING TWENTY AMERICAN APOLLOS AND DI-
ANAS, LAUNCHES NATION-WIDE MOVE TO ENCOURAGE
WEDDING OF PHYSICALLY FIT: IDEA WINS SPONTANEOUS
ENCOURAGEMENT OF CLERGY.

Entrants, required to submit photos of themselves in swim-wear, would be matched with potential mates by Macfadden and the paper's team of experts. Were a couple to marry, they would receive $1,000, plus $100 for each physical-culture baby they produced.

As if he were afraid the contents weren't enough to catch the world's attention, Macfadden opted to print the *Graphic* on shocking pink newsprint. The paper's freak-show nature caused a brief sales spike to four hundred thousand copies per day, which rapidly dropped to below one hundred thousand. Macfadden told Gauvreau that his job was to get the number to five hundred thousand. "With that number of readers, people will listen to us, and we can carry on our campaigns effectively for public health," he said.

Another group dedicated to public health had been keeping a close watch on Macfadden. The American Medical Association, the organized face of the monopoly that to Macfadden represented "allopaths," "the medical trust," "pus-instillers," and "the murderous science" had been absorbing the crusader's abuse publicly for a quarter century. Though it would have been impossible to know from reading *Physical Culture*, the AMA of 1924 was a vastly different organization than it had been at the turn of the century. In 1899, the year *Physical Culture* started, there had been no national standards for medical degrees and licensing. The intervening scientific revolution had transformed doctors from generalists who looked at tongue secretions and felt for weak pulses and warm foreheads to specialists who could employ X-rays and electrocardiographs. A surgeon could remove an appendix ready to burst. Whereas in the late nineteenth century doctors could still be found practicing bloodletting techniques such as leeching and cupping, now a patient's white blood cells could be examined under a microscope. As doctors' knowledge of infectious diseases grew, so did the status of their profession. Medicine had become a serious growth industry, too. America had 42,000 hospital beds in 1909. By 1923 it had more than 750,000.

For a libertarian health pioneer who distrusted doctors, despised drugs, and just couldn't see, as far as the germ theory went, what all the fuss was about, the growth of the AMA into a powerful trade lobby was unbearable. In 1920 a full-page editorial had appeared in *Physical Culture*:

Shall We Have Medical Freedom?

Why Not a National Organization and an Immediate National Convention Including Homeopaths, Osteopaths, Chiropractors, Christian Scientists, Food Therapists, Mechano-Therapists, Hydrotherapists, Naturopaths and other Accredited Schools of Healing That Demand Medical Freedom?

We have religious freedom in this country; we have freedom of speech. But medical freedom we have not.

As the decade moved on, Macfadden's attacks on the AMA intensified. When the New Jersey legislature introduced a bill calling for all natural healers to have their practices examined by AMA-sanctioned physicians, he could hardly control his fury. In a March 1924 editorial titled "Is the Medical Octopus Out to Crush the Nation's Life?" he wrote:

It is a crime for the manufacturers of soap to form a monopoly to control the price of soap, but it is entirely all right for the allopathic school of medicine entirely to control the health and life of the whole of this country according to the American Medical Association. . . . Has not the American Medical Association decided that drugs and serums are the only means of curing diseases and that they have a monopoly on all knowledge appertaining thereto?

As a part of the AMA's major reorganization in 1901, the primary aim of which was to tighten standards, the group established an antiquack squad that became the Propaganda for Reform department in 1906. Its purpose was to investigate claims of fraud in the treatment of patients. Macfadden's name appeared in its files from the first.

By 1924, after years of monitoring its most vocal critic, it is possible that the group began to feel threatened by the potential mass reach of Macfadden's crusading new tabloid. The country was certainly more receptive to the theories he promoted. *Diet and Health, with the Key to the Calories*, by Lulu Hunt Peters, MD, a weight-loss guide that prescribed a regimen very similar to physcultopathic treatment for obesity, was the best-selling nonfiction book of 1924 *and* 1925. Perhaps physicians had grown weary of hearing patients chuckle over the latest antidoctor "true" stories that they saw in Macfadden's publications. It shouldn't be difficult to divine the main thrust of this article from *Physical Culture*:

A Doctor's Wife Speaks
The Story of a Physician's Wife Who Witnessed Her Husband's Downfall from a Conscientious, Though Poorly Paid Practitioner to an Unscrupulous Money Grabbing Panderer to the Imaginary Ills of the Rich

Whatever the last straw was, the sleeping AMA giant decided finally to lash back. The group published two primary magazines at that time. The *Journal of the American Medical Association (JAMA)* was, as it still is, a compendium of medical studies and clinical trials. The other publication, *Hygeia*, was a mass-market monthly that attempted to disseminate health information among a lay audience. Its editor was Dr. Morris Fishbein, who according to the AMA's official history, was in 1924 "beginning a reign of 25 years during which he achieved a power and public stature probably unmatched by any physician in [the twentieth] century." He was as firm in his defense of orthodox medicine as Macfadden was in his defense of physcultopathy. The November 1924 issue of *Hygeia* carried a notice alerting its readers to an important new series, one "discussing the manner in which the desire for health and the hope of relief from suffering and disease are exploited by the promoters of peculiar cults and fads." The series started with a two-part analysis of the work of the AMA's most despised health faddist. "The publication known as *Physical Culture* founded by one Bernarr Macfadden is an outstanding example of the money that is to be made from catering to ignorance and furnishing a contact between the quack and his victims."

The article was penned by Fishbein himself. He started by giving Macfadden credit for offering some commonsense advice on diet and exercise, though he also expressed disappointment that the publisher "uses the erotic appeal in his teachings." Niceties dispensed with, Fishbein unsheathed his scalpel: Macfadden was a hypocrite who claimed to despise doctors but strove to cram as many physicians' bylines into *Physical Culture* as he could. Most of these "doctors" were actually frauds with little or no medical training. Macfadden was the worst kind of quack—one who preyed on those who were healthy but thought they were sick. (In such a hypochondriac army, Upton Sinclair, whom Fishbein singled out for abuse, would have earned at least the rank of two-star general.) Macfadden allied himself with "borderline cultists": osteopaths, chiropractors, naturopaths, vegetarians, colon-flushers, antivaccinationists and antivivisectionists, and followers of "electricity therapist" Albert Abrams, a man who believed that health problems could be diagnosed and treated with crude radiolike devices. Fishbein also took umbrage at Macfadden's refusal to accept the germ theory, and his insistence that diseases such as syphilis could be cured by fasting.

Macfadden would have gladly confessed to all these alleged crimes, but Fishbein accused him of worse. Macfadden, he wrote, while loudly and proudly broadcasting his claim to be the prime mover behind the destruction of the patent-medicine industry, was also accepting advertisements from the manufacturers of Sargol, a toxic elixir with strychnine as one of its active agents, and Absorbine Jr., which in those days was marketed not as a back-pain reliever but the "Only Remedy Known that Positively Cures Varicose Veins." In one instance, a company promising to stretch a person's height by two to five inches was sending out facsimiles of a signed testimonial from Macfadden Publications president Orr J. Elder. In one such letter, reprinted in *Hygeia*,

Elder claimed to personally know many satisfied customers of the skeletal extension course. The nation's leading organ for drugless healing was also accepting ads for correspondence courses in pharmacy.

Physical Culture, which had grown to a thick 150 pages per month, contained dozens of pages of dubious advertisements for miracle cures, self-improvement courses, bodybuilding regimens, and Macfadden's own books. The advertising in the back of the magazine, especially, was a Wild West of alternative therapies. In one advertisement, Charles Atlas endorsed the Battle Creek Health Builder, one of those vibrating-band waist-reducing machines that found their highest calling as comic props in Three Stooges films. Instead of denying the *Hygeia* claims, Macfadden ran his own series painting the AMA as a cabal making secret plans to control the world's health from its fortified new headquarters in Chicago. "The million dollar citadel of the medical trust . . . was so constructed as to withstand attack from any source," reads a photo caption of this medical bunker. "The adjacent buildings on either side are at least one foot away so that no entrance can be effected from one building to the citadel."

How Macfadden rationalized accepting ads that seemed to contradict his physcultopathic principles will probably never be known. The likeliest possibility is that he'd determined that since none of these products and services was medical, then the enemies of his enemy were his friends.

After twenty-five years, *Physical Culture* had matured into Macfadden's pulpit for preaching to the already converted. Readers picked it up for the same reasons people subscribe to political magazines—to hear confirmation of their opinions. The *Graphic* was his new forum to proselytize. In January and February 1925, as almost every newspaper editor in the nation embraced the story of the team of sled dogs racing against time to bring life-saving diphtheria antitoxin to Nome, Alaska—MUSHER WITH LIFE-SAVING SERUM ARRIVES AT NOME was one typical front-page headline, from the *Los Angeles Times*—Gauvreau was running stories purporting to expose the dogsled brigade as a "publicity stunt," and unmasking the hero husky Balto (a statue of whom was unveiled in Central Park the same year) as a fraud. When New York City health commissioner Frank Monaghan earmarked $80,000 in May 1925 to mobilize against a smallpox strain that was spreading up the East Coast, the *Graphic*'s headline was NEW YORK THREATENED WITH SERUM ORGY. After President Coolidge was vaccinated during the same epidemic, the *Graphic* demanded to know IS REPORT OF COOLIDGE'S VACCINATION PUBLICITY STUNT OF PUS TRUST?

After the *Hygeia* articles, the AMA seemed to realize that lavishing negative publicity on Macfadden was like pouring kerosene on a campfire, and adopted a policy of public silence. "Nothing would please Macfadden more than to have the American Medical Association take cognizance of the ridiculous series he is now running on [us]," wrote the head of the AMA's investigative bureau to a concerned physician who'd complained about the *Physical Culture* series. Privately, however, the AMA had the *Hygeia* articles reprinted and sent a free copy to anyone who requested one, as well as to the officials of any town in which they'd learned that Macfadden had scheduled a speaking engagement. In Richmond, Virginia, the AMA's efforts were able to get "all [Macfadden's] lectures . . . promptly canceled," boasted a local doctor. In Buffalo, a "number of prominent physicians" were said to have "threatened to cancel their accounts" at stores where Macfadden had been scheduled to make appearances. Dozens of doctors wrote in, keeping tabs on Macfadden, including his old Battle Creek neighbor John Harvey Kellogg, who reported that from his perch across the street, it appeared that all Macfadden had been doing to treat his patients was to cut off their food.

Macfadden was certain that the medical body had taken even darker measures against him, Mary remembered. "He became convinced the American Medical Association, in order to at least make him sick, had hired spies to poison our wells" at Nyack, she wrote. "While we were confined to buttermilk he sent out

samples of our drinking water to be analyzed." When Macfadden's twenty-one-year-old daughter with Marguerite Kelly, Byrne,
died suddenly of a heart ailment on June 20, 1926, the chief of
the Propaganda for Reform Department wrote to New York City
health authorities "trying to get information" regarding the conditions of her death to "offset this quack's pernicious activities."

Chapter 11

Extra! Extra!

Should I read Euripides or continue with the Graphic?

—COLE PORTER, "WHICH?" 1928

Within days of its debut, the *Graphic* had been tagged with an unshakable nickname, the "*Porno-Graphic*." The New York Public Library banned its pink pages from the Reading Room after only six weeks. *Time* incessantly referred to the paper as a "gum-chewer's sheetlet." Macfadden lost more than half a million dollars on the *Graphic* in the second half of 1924, and another million and a half in 1925. *True Story*, however, covered these deficits and more. In 1925, the magazine's circulation hit 1.5 million, and Macfadden Publications revenues reached $9 million. The owner was bullish on his paper's prospects. "If we can get the losses down to $10,000 a week we'll be sitting pretty," Macfadden cheerily told Gauvreau.

Gauvreau had squeezed an extraordinary staff into his secondhand downtown newsroom. The first thing a daytime visitor would have seen after stepping off the temperamental elevator (which was prone to breaking loose and plunging to the basement) onto the editorial floor was Walter Winchell's rolltop desk overloaded with mail, parcels, and other detritus. Winchell worked nights reporting the *Graphic*'s most popular column,

Your Broadway and Mine. He wrote in a jazz-inflected slanguage of his own invention and peppered his copy with "Winchell-isms," colorful terms describing how a man and woman became "an item" (a romantic couple), progressed to "middle aisling" (getting married), and finally expected a "blessed event" (birth) before eventually "Reno-vating" (divorcing). His Monday column, in which he disgorged a week's worth of juicy gossip, was New York's most widely anticipated shovelful of inside dirt. The *Graphic* was his first newspaper job. "No paper but Macfadden's tabloid could have nursed such a prodigy," Gauvreau later wrote.

At the rear of the city room were the contest editor's desk, the lonely clippings archives (Macfadden broke with newspaper tradition in his refusal to refer to this collection by its standard if unhealthy name, the morgue), and the photographic department, where journalistic history of a sort was shortly to be made. Among the employees who came and went behind the desks at the room's center were cub reporter John Huston, whose mother was a *Graphic* writer. Her son was fired by Gauvreau after he wrote a story identifying an innocent man as a murderer. Huston had somewhat greater success in Hollywood, as director of such films as *The African Queen*. Another ultimately westbound protégé of Huston's mother was the future director Samuel Fuller, whom Gauvreau hired as a crime reporter while he was still in high school. (His 1952 classic *Park Row* was inspired in part by his time at the *Graphic*.) Buzzing amid everything was a copyboy named Robert Harrison, who made little impact on the *Graphic* but who absorbed its keyhole-journalism ethos better than anyone save Winchell. He would go on to found *Confidential* magazine, the outrageously successful Hollywood gossip publication of the 1950s that Tom Wolfe called "the most scandalous scandal magazine in the history of the world." Congressman Fiorello LaGuardia, who would soon be elected mayor of New York, wrote a column called I'm Telling You Confidentially! in which he beat the drum for liberal ideals. He once passed an evening with Gauvreau, sniffing perfumed inks that Macfadden wanted to test in his printing presses.

In a narrow room off to one side, the sports department made its home. Here, the dapper Ed Sullivan sat slouched in his chair, shirt cuffs rolled, shooting the breeze with his coworkers. Sullivan, only twenty-five, had been a reporter since his teens, and had the writing of his free-form column, Ed Sullivan's Sports Whirl, down to a science. Prose "cascaded freely right into his typewriter" according to a colleague. Sullivan's list of contacts included Jack Dempsey and Babe Ruth, and he spent his free time organizing sports banquets where he honed the skills that would make him America's prime-time television master of ceremonies for almost two decades. Sullivan also earned extra money—and extra credit with Macfadden—by organizing strongman competitions.

The newsroom nexus, Gauvreau's office, was a shrine to Napoléon, stuffed with books and paintings celebrating the emperor's triumphs. (Some staffers thought Gauvreau even styled his hair in tribute to his idol.) A bust of the Little Corporal stood watch against Gauvreau's most loathed intruders, Winchell and Oursler. The editor got along well with the owner, however, and Macfadden would often walk nearly half the length of Manhattan Island to discuss circulation at Gauvreau's desk and share a bag of walnuts, which Macfadden cracked with his hands. Gauvreau was fascinated by Macfadden—his superhuman energy, his way of cutting through bullshit to find the grain of truth in almost any business matter, his uncanny ability to look a man over once and then predict his life span. He'd also developed a grudging respect for the owner's news instincts. Mary Macfadden had been ordered by Bernarr to keep a pen and paper tethered to their bed in Nyack, to record her husband's middle-of-the-night inspirations for his big pink baby. At three o'clock one morning, Gauvreau was awakened by a call from the boss demanding that news of the Sacco and Vanzetti executions run on the front page under the headline ROASTED ALIVE! That day's paper saw a boost of thirty thousand in sales.

If Macfadden was willing to look the other way in *Physical Culture* when it came to advertisers, he was unwilling to do so in the *Graphic*. Cigarette ads, a staple source of revenue for all other newspapers, were off-limits. (A sign hung in the newsroom: "Please Do Not Smoke. It Raises Our Hazard and Clashes with Our Ideals.") When his salesmen landed a schedule of shoe advertisements, Macfadden responded with an editorial condemning high heels. A hat manufacturer's campaign promoting its new line of men's headwear provoked Macfadden to urge his male readers to liberate their scalps and "air their hair," Gauvreau remembered. The editor's own black locks turned prematurely gray during this period.

In the introduction to his 1931 novel, *Hot News*, a very thinly disguised tale of his *Graphic* experiences, Gauvreau stated that he'd written the book "to reflect on an era of mad journalism which we will never see again . . . as one of the pioneers [of tabloid journalism], I was swept away in the hectic struggle for circulation." Once the paper was up and running, Gauvreau was left largely unchaperoned. Oursler, whose influence Gauvreau had strived to limit, had suffered a midlife crisis, obtained a Mexican divorce, and run off to Europe with the *Graphic*'s Miss Courtesy, an aspiring writer named Grace Perkins.

After a brief dip in 1924, the stock market was on its way to acrophobic new heights. As securities prices soared, Americans fell victim to a collective giddiness. Coast-to-coast broadcasting and news services had made instant national fads and celebrities possible. Thanks to tabloid culture, more and more of that news centered on figures like the flagpole-sitting champion Alvin "Shipwreck" Kelly. New York was emerging as the media capital of the world; its newspaper editors decided what did and didn't deserve to be called news in the rest of the country. To a large degree, the *Graphic*'s multiple daily editions set the pace—and tone—for news gathering in the Big Apple.

With Oursler completely out of the picture, Gauvreau was free to fully indulge every salacious impulse he'd thus far repressed. The *Graphic* kicked off the greatest eighteen-month party in tabloid journalism history in November 1925, with the surprise marriage and bizarre divorce proceedings of Alice and Kip Rhinelander.

Kip Rhinelander was the slightly dumpy scion of one of New York's wealthiest families, who had fallen madly in love with, and married, the laundress Alice Jones. Within a month, however, Rhinelander's blue-blood family discovered that Alice's father was black. Kip was disowned by the Rhinelander clan, upon which he claimed, "If I ever suspected that there was a drop of colored blood in that woman's veins, I'd never have married her." Alice's lawyer had a plan to prove his claim impossible. He would have his client undress in court and reveal portions of her body untouched by the sun—but presumably viewed by her husband in their conjugal bed—which would advertise her heritage beyond the shadow of a doubt.

The entire *Graphic* staff knew that Macfadden wanted as much skin in his newspaper as he could get away with showing. When the art department received a terse memo from uptown that read "Put passion in fashion," they responded with a photo illustration peeling away the layers of a career girl's office outfit until only her bra, panties, and garters remained. Not all flesh was up to Macfadden's connoisseur standards, though. The editor who ran a picture of Marlene Dietrich received a tongue-lashing from the boss: "I don't want any more skinny-legged women in this newspaper. I want women with hips and breasts."

Confronted with the Rhinelander story, the *Graphic*'s assistant art director, Harry Grogin, had a brilliant idea. Since no cameras were allowed *in* the courtroom, why not photograph a naked woman and superimpose her onto a picture *of* the courtroom? Twenty separate photos, including one of a nude showgirl who posed topless for the paper (photographed from behind; her skin was lightly tinted for a mulatto effect) were cut and pasted into a brilliant montage that, unless one looked closely, could have passed as a picture of the actual unveiling. ALICE DISROBES IN COURT TO KEEP HER HUSBAND was the headline, printed in a size other papers usually reserved for declarations of war. A minuscule notice in the caption acknowledged that the photo was a composite. Gauvreau's doppelganger editor in *Hot News*, who invents a similar mongrel process for his tabloid, admits that "most people bought the paper under the impression that it was a photograph."

Each time the *Graphic* ran a montage of this sort—the steamier, the better—the paper's circulation spiked from an average of sixty thousand per edition to several hundred thousand copies. Gauvreau, delighted, gave the composite photos a new name, composographs, which he had trademarked in Washington, D.C.

As the spring of 1926 approached, the *Graphic*'s thoughts lightly turned to warped love. Edward "Daddy" Browning, a middle-aged real-estate tycoon with what one *Graphic* editor called "a weakness for young girls and publicity," began courting a chubby high schooler named Frances Heenan, who would forever be known by the nickname Peaches. GIRL, 15, HOPES TO MARRY BROWNING, 58, was a banner *Graphic* headline on April Fools Day. They wed two weeks later, with her mother's blessing.

Immediately thereafter, Peaches's serialized "honeymoon diary" began running in the *Graphic*. The stories about the newlyweds were inane, but the tongue-in-cheek composographs verged on genius. Scenes were invented of Daddy taunting Peaches in their bedroom with a doll (a reference to his refusal to have children); Daddy dressed as a Valentino-inspired caliph to amuse his wife; Daddy romancing Peaches on a settee as her mother ominously eavesdropped at the door. Gauvreau persuaded Browning to adopt what he assured the millionaire was a "rare African honking gander," and the bird appeared in the composographs as well, furthering the merriment by offering editorial comment in cartoon bubbles. Never one to miss out on good publicity, Macfadden challenged Browning to a spinach-eating contest. The winner was not recorded for posterity.

Life at the *Graphic* offices was often as strange as the world it covered. Long after the paper's self-help novelty had abated, the parade of strongmen and health nuts continued through its aisles. Reporters had learned to squirrel away telephone directories from hypertrophied would-be show-offs, but now the *Graphic*, which regularly published tales with titles like MY JAW STRONGEST IN WORLD was a beacon to those who wanted an audience as they pounded nails through wood planks with their fists or hung from ceiling pipes by their hair. Ed Sullivan was once hounded into following a behemoth over to Manhattan's West Side Railyards to watch him push a train with his head.

Macfadden sometimes arrived at the offices late in the morning, invigorated by the twenty-five-mile walk he'd made barefoot from Nyack, his head filled with ideas. One day he might write an editorial stumping for colonic hygiene, the next he might order a crossword puzzle contest with $25,000 in prizes. When a former junkie praised Macfadden for curing his addiction through physcultopathic methods, the *Graphic* ran a notice throwing open the doors of the newsroom to any other lost souls who might wish to hear about Macfadden's miracle treatment. "That evening our offices were turned into a raving asylum by all manner of terrifying, shaking creatures who crawled with pawing hands over shuddering copyreaders, climbed on chairs and finally had complete possession of the place," Gauvreau recalled.

Only a blockbuster story could have knocked Daddy and Peaches off the *Graphic*'s front page. In July 1926, one did. The Reverend Edward Hall and his lover, Eleanor Mills, had been murdered in 1922. Mills was a singer in Reverend Hall's choir in New Brunswick, New Jersey. Both were married; Hall's wife came from one of the most powerful families in New Brunswick, which may have accounted for the absence of arrests when the intertwined bodies of Hall and Mills were discovered beneath a crab apple tree, love letters scattered across their corpses. The case remained dormant until Hearst's *Daily Mirror* reopened it with fresh evidence. At least some of this proof had apparently been unearthed by the *Graphic*. In the frenzied months prior to the paper's debut, Gauvreau had assigned a reporter to investigate the unsolved murder in hopes of landing the scoop of the century. Uncharacteristically, Macfadden had nixed the story as too flimsy.*

In August the defense and prosecution hunkered down to prepare for the Hall-Mills trial. During the first three weeks of what is typically the calendar's sleepiest month for news, the *Graphic* and other tabloids turned their gaze away from New Jersey only twice. On August 7, front pages blared the news that the daughter of an Upper West Side butcher, the nineteen-year-old *Physical Culture* pinup Gertrude Ederle, had become the first woman to swim the English Channel, in fourteen hours and thirty-one minutes. Two weeks later, composer Irving Berlin and his wife, the heiress Ellin Mackay, who had escaped to Europe after secretly marrying in January against her father's wishes, were mobbed by reporters when they returned to the city.

Gauvreau was among those who'd fled Manhattan for the month. He'd embarked on an extended trip south, sailing to Nicaragua and Honduras on a fruit steamer to interview the charismatic leader of a Central American rebel army. En route, the boat was rocked by one of the twentieth century's most vicious hurricanes. If the *Graphic* editor, facing death at sea, repented for his sensational ways, it's possible that he might have returned home and shaken his paper, and its owner, out of their tabloid delirium.

When the Mirror's *story broke after midnight, the* Graphic's *night editor hastily sent the only two staffers still in the office, a pair of young reporters, out to New Jersey to dig up whatever facts they could. Had the car carrying Walter Winchell and Ed Sullivan met ill fortune that night, American gossip would never have been the same.*

In New York, however, events were unfolding that made such a conversion impossible. Rudolph Valentino, the thirty-one-year-old screen star and *Physical Culture* model, had been rushed to Manhattan's Polyclinic Hospital for treatment of appendicitis and gastric ulcer. The *Graphic* announced his doctors' pessimistic prognosis for recovery with the headline: VALENTINO DEAD, beneath which appeared the tiny clarification, "Broadway Hears." Just after noon on Monday, August 23, his death became official. For Macfadden, this was a five-alarm story: Not only was the Italian a huge celebrity, Macfadden had printed a book about Valentino's diet and exercise regimen, titled *How You Can Keep Fit*. Even before prematurely announcing that the Sheik had passed on, his suffering had inspired a composograph that bore Macfadden's unmistakable touch: the actor lying on a hospital gurney, surrounded by surgeons ready to commit unnatural crimes against his beautiful physique: RUDY BRAVE IN THE FACE OF DEATH was the headline. Another faked exclusive, an image of the actor lying in state, kept the *Graphic*'s ancient presses running through the night. The next day's front-page image was even better: Valentino being welcomed into the kingdom of heaven.

Thirty thousand grieving fans formed a line eleven blocks long to pay their last respects to Valentino. As the crowd thickened, some feared they'd miss their chance to view the body and so began pushing and shoving. A riot erupted. Windows were smashed, dozens of people were injured, and thanks largely to every newspaper's attempts to keep pace with the *Graphic*, a one-day story stretched more than a week through the late-August doldrums. The *Graphic* churned out conspiracy theories: Rudy had been poisoned; Rudy had been thrashed to death; Rudy had been secretly gunned down in a restaurant. (Macfadden, under counsel from his libel lawyer, held his tongue in the paper, though he slipped an article revealing "What Really Killed Valentino" into *Physical Culture*. Not surprisingly, doctors were the culprits.) *Graphic* sales soared by more than a hundred thousand copies per day. On the day of the funeral at St. Malachy's church on Forty-ninth Street, the *Graphic* was hawking a special issue with a front-page composograph of the burial procession *before* the ceremony began, making it possible for mourners arriving at the service to experience a moment of tabloid déjà vu. The entire print run sold out.

On its best days, sales of the *Graphic* were now consistently exceeding Macfadden's goal of five hundred thousand copies. On days when those numbers plummeted, the owner took consolation in the cavalcade of fine physical specimens who paraded across his front page in the autumn of 1926. The Gene Tunney–Jack Dempsey heavyweight bout provided an excuse to run a week's worth of photos of muscled, shirtless men. In late September, Macfadden's former physcultopathy student Amelia (Millie) Gade Corson bested Gertie Ederle's record for swimming the English Channel. In a column marking the *Graphic*'s second anniversary, Macfadden discussed the results of a reader survey that seemed to affirm his aspiration for the paper. "More than a third of you prefer the Physical Culture page to any other feature. . . . The *New York Times* is your favorite among [other] morning papers, then along come the *World* and the *Herald Tribune* and the *American* . . . it looks as if readers of THE GRAPHIC prefer the BEST." He celebrated by purchasing a $4 million eight-story building under construction on Hudson Street as the new headquarters for his tabloid.

Having seen upon his return from Honduras how the Valentino story had temporarily pushed the *Graphic*'s circulation into profitable territory, Gauvreau issued a memorandum to the staff that read, in part:

I want the reporters to bear in mind that from now on it is a waste of time to write a story that cannot stand up under a sensational hed.

Always keep the Kip Rhinelander picture in mind. This was strikingly sensational.

Handle every item from the sensational angle.

If a story cannot give you an interesting, and sensational head-line, it has no place in this paper from now on.

As if on cue, the tabloid's favorite players reentered the scene. The Hall-Mills trial began, with four of Reverend Hall's wealthy in-laws accused of murder. The case's first star emerged almost immediately: Jane Gibson, who owned the hog farm where the bodies had been found. Gibson's story was that she and her mule, Jenny, had been walking near the murder tree that night; she'd heard someone say, "explain these letters"; then she'd heard gunshots. As she was wheeled from the courtroom she turned to the defendants and screamed, "I have told the truth! So help me God! And you know I've told the truth!" (Because Hearst's *Mirror* owned this story, the *Graphic* played up doubts about the guilt of Hall's relatives; PIG WOMAN'S MOTHER SAYS DAUGHTER NOT IN DE RUSSEY'S LANE THAT NIGHT, was its headline about Gibson's testimony.) After months of coverage, and millions of papers sold, Gauvreau finally blocked out an afternoon front page that read FOUR HALL KIN FREED, ALL IN-DICTMENTS QUASHED.

Conveniently, Peaches Browning had just walked out on Daddy. The *Graphic* paid Peaches $1,000 to tell her side of the tale, which was counterweighted with the journalistic balance of stories such as GRIEF-STUNNED BROWNING AT FLIGHT OF PEACHES FROM THEIR LOVE NEST "by Edward West Browning." When the inevitable divorce proceedings followed, Gauvreau as-signed two reporters full time to Peaches and another to Daddy. For weeks, little else mattered as far as the paper was concerned. The Brownings were circulation gold: PEACHES ON STAND TELLS HOW DADDY MADE LOVE was a typical headline. Details spilled out from the courtroom on a daily basis: Daddy's insistence that Peaches parade nude around the house; her mother's assistance in birth control; the missing pages from Peaches's diary, which con-tained tales of her experience with the opposite sex prior to meeting

Daddy: PEACHES ADMITS HIDING NAMES OF BOY LOVERS . . .
GIRL QUIZZED ON MARITAL DEPRAVITY.

Macfadden was ecstatic over the newsstand numbers that
were rolling in. Sales of editions bearing the most preposterous
composographs had soared above six hundred thousand, and
readers were hooked on the wink-wink coverage. Winchell's star
had continued to rise. (To Macfadden, though, as Gauvreau
noted, Winchell's column was "a mass of unintelligible jargon";
the boss greatly preferred comic strips such as *L'il Samson*, in
which the hero invariably encountered combination safes falling
from the sky and movers perplexed by unbudging grand pianos.)
Uptown, the *Daily News* had grown into the nation's largest-selling
paper. The *Graphic* battled with Hearst's *Mirror* for second
place, and survival. When bound piles of the Sunday *Graphic*
were spotted floating down the East River, Gauvreau suspected
Hearst. The editor retaliated by printing innocuous stories from
the life of Marion Davies, Hearst's live-in girlfriend, in a menacing typeface. The thefts came to an abrupt end.

Not everyone was amused by the *Graphic*. New York's advertising community still avoided the paper. Civic leaders in towns
outside of New York's five boroughs, incensed by the *Graphic*'s
coverage of the Browning case, were threatening to shut down
newsdealers who carried the tabloid. Vice squad leader John S.
Sumner was still busy making life difficult for art dealers, bur-
lesque show producers, and newspaper publishers. On February 5,
1927, Macfadden and Gauvreau were summoned downtown
to the Tombs Court to face inquiries from Sumner "regarding
certain objectionable matter that has been appearing in the
Graphic," according to the *New York Times*. "The handling of
the Browning separation suit and the accompanying composographs," Sumner said, were "the chief cause of the many complaints he had received." Sumner told *Editor and Publisher* that
"The *Graphic* has acted like a small child trying to be nasty. Its
stories on the Browning case have been simple, inane and foolish,
with a strain of dangerous filth running through them harmful to

all society." Macfadden sought the counsel of the best defense lawyer he could think of, William Travers Jerome, a onetime anti-corruption district attorney whose best years were long behind him. Macfadden and Gauvreau paid a visit to Jerome, who, it turned out, had never heard of the *Graphic*. Gauvreau produced a copy, which Jerome scanned quickly and handed back.

"Plead guilty," he said.

A new lawyer was found, and the case was eventually dismissed.

On Thursday, March 17, 1927, Rudolph Valentino made a miraculous comeback on the front page of the *Graphic*, in a composograph that may have been the finest ever created. RUDY MEETS CARUSO—TENOR'S SPIRIT SPEAKS! read the enormous headline, which ran across an image of the late operatic superstar Enrico Caruso welcoming his countryman into heaven, "soon after reaching the great beyond." The new pals looked beatifically content dressed in their long, white robes. The source for the story was unimpeachable: Valentino's wife, who had received it courtesy of her psychic.

On March 21, all of New York was awaiting the following morning's verdict in the Browning divorce trial, when tabloid pandemonium struck yet again. Albert Snyder, a magazine art director, was found murdered in his Long Island home. While asleep, Snyder had been chloroformed, bound hand and foot, bludgeoned, and strangled. His wife, Ruth, claimed a burglar had attacked the couple. Police found the murder weapon and a tie pin belonging to Ruth's lover, a corset salesman named Judd Gray. Both killers quickly confessed, then blamed each other.

Snyder-Gray was *Graphic* catnip. Ruth was an exotic beauty, and obviously had mesmerized Judd with her sexual allure; he referred to her as Momsie in his love letters. Judd claimed that Ruth had attempted to kill Albert seven previous times, once putting poison in his prune whip. (Albert, for his part, was not the most sympathetic of murder victims. He had been obsessed with his dead fiancée, whose cherished portrait he hung in a place of honor in the home he shared with Ruth.) Macfadden wanted Gauvreau to underscore that this crime had been committed by persons indulging in bootleg alcohol and nonprocreative sex. Gauvreau's choice of a writer to cover the trial was inspired: Aimee Semple McPherson, the Southern California revivalist minister who packed fifteen thousand faithful into her Angelus Temple each day and reached millions of others by radio.

Huge crowds clamored to get inside the packed courtroom to see what was, now that a few months had passed since Hall-Mills, the new Trial of the Century. During the afternoon of May 3, in what may have been the zenith of Jazz Age tabloid journalism, Gray was called to testify. He recounted the night of the murder so vividly that Albert Snyder's brother leaped from his seat screaming, "Make him stop! For God's sake, make him stop!" At the instant Gray was describing the kill, Ruth leaped screaming from *her* chair, lunging at the witness box. Snyder's brother collapsed; Ruth was restrained; Gray fainted in his seat; the judge adjourned for the day. Newsmen scrambled to file copy for the afternoon final editions, and presumably toasted their good fortune into the wee hours. GRAY'S STORY SEARS RUTH was the *Graphic*'s headline.

The day the verdict was due, Macfadden personally dictated a banner headline combining two monumental events expected to coincide:

FRENCH FLIERS LAND IN TIME TO HEAR SNYDER VERDICT

All through the spring, a craze for transatlantic flight had been building. Though the ocean had actually been crossed by air in 1919, from Newfoundland to Ireland, airmen were now chasing the Orteig Prize, which promised $25,000 to the first pilot to fly nonstop in either direction between New York City and Paris. The French World War I heroes Charles Nungesser and François Coli had departed Paris at dawn on May 8 to fly westward. The next day, French newspapers too eager to wait for facts to arrive invented details of the pair's triumphant arrival over Manhattan, including a salutary spin around the Statue of Liberty.

The reports were wrong. Nungesser and Coli were never seen again. Their disappearance only heightened the tension in New York, where reporters hounded three fliers who were ready to attempt the journey from the opposite side of the pond. One of them, a slim, handsome Minnesota lad named Charles Lindbergh, got the jump on his rivals on May 20. Thirty-three hours and twenty-nine minutes later, he arrived at Le Bourget Airfield. The front page of the *Graphic*'s special edition, printed while the pilot was still winging over the British Isles, was uncharacteristically subdued. A photo of a smiling Lindbergh standing aside his plane ran under the headline WELL, I MADE IT.

Though Macfadden didn't know it at the time, tabloid fever had broken.

Chapter 12

Fame, Fortune, Fascism

*I began to have nightmares about what might happen if the
Men of Muscle took over the White House.*

—MARY MACFADDEN, *DUMBBELLS AND CARROT STRIPS*

As the *Graphic* surged ahead, blazing trails of bad taste, the
rest of the Macfadden Publications empire thrived financially. Even with the anchor of the *Graphic*'s start-up costs weighing on its earnings, the company's stock price had risen more
than 1,200 percent since shares were first offered in 1919. Macfadden may have been a shoeless health fanatic with rumpled
suits and a Swiss-cheese hat, but no one could doubt that he also
possessed an uncanny sense of what the public wanted. As often
happens in times of great prosperity, businessmen in the
mid-1920s were celebrated for their brilliant management. Macfadden was more eager than most to accept his financial success
as evidence of his leadership qualities.

Over the years, as his magazines' reach had extended exponentially, Macfadden had become an unavoidable presence in
their pages, through editorials boosting his prohealth themes as
well as more aw-shucks advice about the virtues of hearth and
home. Subtly, *True Story* had moved away from sex toward romance. In 1924, ever sensitive to legal trouble, Macfadden had

suggested bulletproofing the magazine against growing moral criticism by hiring a ministerial board. The committee assembled by Oursler at his boss's behest sounds like the setup to a vaudeville joke: ministers from the Methodist, Presbyterian, and Congregationalist faiths, as well as a rabbi, were each sent copies of stories that had been accepted for publication in *True Story*. (A Catholic priest dropped out after his bishop disciplined him.) If any of the spiritual advisors objected to a word, it was changed. If one found a story to be inappropriate, it went unpublished. Sales of the clerically approved *True Story* continued to rise. In its seventh year, monthly circulation topped 2 million.

True Story was still downmarket, however. One social worker interviewed for a study noted that single mothers "almost universally came to the maternity home armed with *True Story* magazine." Oursler moved William Jourdan Rapp over from *True Romances* as *True Story*'s new editor in 1926, with the mandate to remake it into a mainstream women's magazine, like the Big Six, and to burnish its appeal as an advertising vehicle. He succeeded so well that he kept the job for sixteen years. Much of *True Story*'s male-oriented content had already been siphoned off into *True Detective Mysteries*. The net financial result was a plus—*True Story* cleared $2 million in 1927 and $2.5 million in 1928, and *True Detective* was profitable almost from the first issue.

True Story wasn't the only magazine on Sixty-fourth Street keeping its pants on. Beginning around 1927, *all* Macfadden publications outside of the *Graphic* dialed back their sex quotient. *Physical Culture*, which for most of the decade had been jammed with photos of America's finest physical specimens—not a few of them tastefully posed, liberally greased nudes—was publishing noticeably fewer pictures of "full-figured women [and] brawny near-nuded men with marcelled hair and muscle-bound expressions," said *Time*. New Macfadden magazines were aimed at the prosperous and growing suburban audience. *Your Home* was published for middle-income couples who planned to have a house built especially for them. *Your Car* was sort of an upscale

True Story in which all the action unfolded within fifty feet of a steering wheel.

A likely cause of this giant step away from sex was Macfadden's new, serious aspirations. From its start, the *Graphic* staff had suspected that the paper's unspoken purpose was to give Macfadden a populist forum to establish a political career. In his memoirs, Oursler wryly called the *Graphic* Macfadden's "magic carpet that would carry him to the White House." Much of the staff, and certainly Mary, had no doubts that Oursler was the genie who'd conjured for Bernarr Macfadden the vision of himself as a public servant. While in the 1920s the American public probably needed a few more decades to acclimate itself to the idea of musclemen strolling the corridors of power, it isn't surprising that Macfadden saw the shift to politics as a natural one. Fiorello LaGuardia wrote for the *Graphic*. Royal Copeland, one of New York's two senators, was a *Physical Culture* contributor and licensed homeopath (as well as a renegade MD for whom the AMA's Morris Fishbein had nothing but contempt). Upton Sinclair had made two bids for Congress, as a socialist no less. And William Randolph Hearst had parlayed the pull of his yellow-journalism newspapers into two terms in the House of Representatives.

In 1925, Macfadden made sure that the paper supported playboy Jimmy Walker's candidacy for mayor, with the implication that Walker might appoint him as the city's Commissioner of Health—the position Royal Copeland had held prior to joining the Senate. Walker didn't. "Everybody knows you can live to be a hundred by following Macfadden's ideas," the new mayor quipped. "But New York wants to live the way I do." Mary said her husband reacted to the bad news by going on a week's fast.

As his political ambitions grew, Macfadden's hunger for publicity soon rivaled Daddy Browning's. As often as not, he held his family up as evidence of his leadership and efficiency. The births of his sons were trumpeted far and wide as proof that he could solve seemingly impossible problems. He kept a four-inch-thick binder of press clippings about his daughters, which he flipped through constantly. The *Graphic* itself was treated as something of a family scrapbook, showcasing photos of the children dancing in classical costumes or seated alongside movie stars and sports celebrities; one editor said that the scandalous tabloid was in fact a family newspaper—the Macfadden family newspaper. Because they were the manifestation of all their father's theories, the young Macfaddens were kept on a strict exercise regimen. Cornelius Vanderbilt Jr., the Manhattan heir and a regular Macfadden magazines contributor, recalled a visit to the Nyack house:

The family—Mr. and Mrs. Macfadden and seven or eight daughters—retired early, but I was warned that a bell would ring around 6:30 AM for a dip in the swimming pool. It was a cool evening and an especially cool morning after, and when the bell rang I wasn't in the least interested in going for a swim. But, after all, I was a guest, and I was interested in a [writing] contract, so it seemed best to appear.

When I looked out my window I was amazed and a little shocked to see Mrs. Macfadden and the children, ranging in age from perhaps five or six up to fifteen or sixteen, and also their French governess, all lined up around the pool in their scantiest attire. In those days the Bikini bathing suit was unknown, but they were wearing something like it—a sort of jock strap around the middle, and almost nothing above the waist. Then I saw Mr. Macfadden go out, with all his muscles showing; he was a tremendously well-built man. Someone blew a bugle and they all started marching around the pool. Then they began doing somersaults and calisthenics, and next they began diving into the pool, although the temperature couldn't have been much over fifty degrees.

Macfadden's closest advisors persuaded him to hire the public-relations genius Edward Bernays as a consultant. Bernays had succeeded in making the stone-faced Calvin Coolidge seem almost approachable before his landslide victory in the 1924 presidential election. In 1927 he was working on his book *Propaganda*, which Noam Chomsky has described as "the main manual of the public relations industry." Bernays's assignment, he said later, was to "untangle Macfadden's socially constructive ideas from fads and cults." The argument that serves as the foundation of *Propaganda* indicates that he must have found his new client a fascinating case study as well: "In almost every act of our daily lives, whether in the sphere of politics or business, in our social conduct or our ethical thinking, we are dominated by the relatively small number of persons . . . who understand the mental processes and social patterns of the masses. It is they who pull the wires which control the public mind."

Once a week, Bernays would get together with Macfadden, Gauvreau, Elder, and other top Macfadden Publications executives. Macfadden "sat at these meetings straight as a poker on a stenographer's chair," Bernays recalled, and usually said little except to toss out the occasional outlandish idea, such as suggesting that he walk barefoot from Sixty-fourth Street to the south end of Manhattan to "demonstrate how healthy living preserved the vigor" of a fifty-nine-year-old man. Another time he proposed that a statue made of his daughter Byrnece when she was twelve years old—a statue of his *nude* twelve-year-old daughter, that is—"be sent on a tour through the country to demonstrate what his adherence to physical training habits had done for his offspring." Bernays was able to quash both ideas, but not Macfadden's paternal pride in his pubescent daughter's physique. "Byrnece had a degree of physical development which should be attained by all girls who are to be future mothers of the race," he told one of his authorized biographers a year later. "This replica of her body will help keep such an ideal alive." Byrnece's likeness remained where it was, standing on one side of a giant fireplace in the Nyack home; on the opposite was one made of her father by the noted sculptor George Gray Bernard.

Bernays focused on making Macfadden's public image less health-obsessed and more august and statesmanlike. Testimonial lunches and dinners, attended by Mayor Walker and other dignitaries, were organized to honor Macfadden's achievements as a health pioneer. (Macfadden Publications always picked up the tab.) Bernays arranged for Macfadden to visit England for two weeks. Reporters recorded Macfadden's first meeting with *Physical Culture* contributor, fellow antivaccination enthusiast, and longtime pen pal George Bernard Shaw, for whom Macfadden proudly screened *Rampant Youth at Sixty*, a movie short Pathé Films had shot of him working out. Bernays fixed up a testimonial dinner for Macfadden at the House of Commons, where the honored guest told the assembled legislators that "Britain would gain, and not suffer, if you copied Germany and created a minister of athletics." Back in New York, Gauvreau was ordered to create a special four-page booklet commemorating this momentous event. He chose to pose his boss opposite Daniel Webster, who had made a similar journey in 1839. This "unique treatise" was then "mailed by the thousands to members of every legislature in the country, countless clergymen and Washington lawmakers."

Upon his return to New York City, Macfadden and family were whisked via motorcade to City Hall for a welcome-home reception emceed by Mayor Walker. Macfadden then sped off to Washington, D.C., where Fiorello LaGuardia saluted his perfect physique as he introduced him at a luncheon for a hundred congressmen. Major papers nationwide carried the news that Macfadden's life had been insured for $1 million, believed to be the highest amount ever underwritten for a man of his age. Bernays received a stupendous salary of $500 a week for his work—almost as much as Oursler was earning—and the headlines that Macfadden pulled during that time indicate that he was worth the expense. Evidently it wasn't enough, and the two men parted ways after a year.

Macfadden was agnostic about party affiliation; what he cared about most was who could secure him a position in power. Republican National Committee chairman Will Hays visited the home in Nyack and came away with a $10,000 donation. Almost simultaneously, Macfadden dispatched Gauvreau to Tammany Hall to find out how much it would cost to get the Democratic nomination for governor, since the popular Al Smith was vacating Albany to run for president. The unofficial boss of the city's political machine, Judge George W. Olvany, dissuaded Gauvreau. "Macfadden would be a popular candidate," he told the editor. "But the propaganda that would be brought against him by the medical profession of the nation would prevent his chance of election. The doctors will never forgive him."

In the summer of 1928, Macfadden bought "one of the swiftest and most luxuriously appointed cabin airplanes ever constructed" and had himself flown to the Republican National Convention, which commenced on June 10 in Kansas City. He wired back to the *Graphic* what he considered to be the most urgent news of the gathering. "Failure of the Republicans to adopt my suggestion for endorsement of the proposal to create a new cabinet office—that of secretary of health—and to make it plain that this official should be non-medical, may rebound to cause trouble," he wrote. "It is entirely possible that the Democrats will not make such a mistake."

After Herbert Hoover was nominated by the Republicans, Macfadden flew to D.C. to congratulate him. He then departed for the Democratic Convention in Houston, though not before stopping to outfit his plane with a small office, an aerial camera, photographic lab, and traveling secretary, so that he might provide panoramic coverage of the event for the *Graphic*. He paused en route to Houston in order to purchase a second plane, for $30,000, which he planned to rename *True Story* and enter in a race from Long Island to Los Angeles. The morning after payment, his pilots wrecked the plane in Indiana. He bought another plane, and another, and another.

Near the end of the 1920s, with Macfadden Publications stock at its height, Mary estimated her family's fortune to be worth $30 million. Macfadden had earned $1.8 million in salary and dividends in 1928 alone. After years of abstemiousness, he began to spend this capital almost as quickly as it came in. He was sponsoring the *True Story Hour* and *Physical Culture Hour*, both broadcast nationally each week. A *True Story* film division had been started, producing at least eight feature movies (all sadly lost); surviving production notes indicate that morals were affirmed and much scenery was Fletcherized. While flying around the country, he found time to purchase the Castle Heights Military Academy in Lebanon, Tennessee. This sprawling 225-acre campus in the foothills of the Cumberland Mountains would serve as "a laboratory for testing out his theories in body-building among schoolboys." Macfadden immediately upgraded the athletic facilities—henceforth every student was required to participate in some sport every day—and overhauled the cadets' diet. Meat consumption was slashed, a daily green salad was mandated for every boy, and such a huge volume of milk was served that the school operated its own dairy. Macfadden was so certain that his Castle Heights boys were in superior shape after only six months that he ordered Gauvreau to have Ed Sullivan set up a football game between his team and "the toughest, hardest, and best school team in New York." Sullivan selected the perennial powerhouse St. John's Preparatory in Brooklyn, and Macfadden brought his boys to the big city. The game, played in a snowstorm at Ebbets Field, was a rout. The squad from Castle Heights shut out the metropolitans, 25–0.

With Byrne's death, the physical culture family was reduced to its final roster of ten members, including a new baby boy. On the night of the *Graphic*'s last test run in 1924, Macfadden had returned home and announced to Mary that such a grand occasion was an auspicious moment to conceive a son. He was right. Mary wanted to name the boy Bruce; Macfadden informed her via office memo that the son would be named Brewster, which he believed conjured up the toughness of a fighting cock.

The larger Macfadden Publications grew, the less room there seemed to be in it for Mary. Macfadden's loyal acolyte Susie

Wood was recalled from Chicago to work in the Sixty-fourth Street headquarters. Wood was soon installed in one of the two offices that adjoined Macfadden's; Oursler still occupied the other. Mary had mixed feelings about Wood's long history with her husband. Wood had been his most devoted follower going back to the dark days of the Physical Culture City fiasco. However, Mary observed later, "whether she confined herself to cracked wheat in those days was to remain a matter beyond my speculation." Mary, who had continued to think of herself as her husband's full and equal business partner, learned that Wood had been appointed treasurer of the company when she saw Wood's signature on a stock-dividend check. Macfadden asked his wife to explain to Wood her duties as head of the manuscript-reading department. Not long after, a memo was circulated informing all editors that "all matters pertaining to manuscripts" must be routed through Mrs. Wood's office. Soon enough, Mary had to inquire with Wood when visiting the office if her husband was free to see her. By 1927, Wood was so occupied with helping Macfadden arrange his ambitious plans that she asked Mary to take care of her nine-year-old daughter, Ruth, for a year.

With the patriarch's sixtieth birthday approaching, all the Macfaddens loaded onto the private rail car *Palm Beach* and set off on a photo-op tour of the life and times of Bernarr Macfadden. The family made well-publicized stops for Macfadden to address Rotary Clubs and other groups of businessmen, then continued on to Castle Heights Academy and St. Louis. Here he gathered his extended family of Millers and McFaddens for a celebratory dinner. Later, the caravan moved on to a stop near his hometown of Mill Spring, where Macfadden held a barbecue for 1,400 locals and showed his children the tiny farm where he had been born in 1868. Just in case they weren't impressed, Macfadden scheduled a stop in Macomb, Illinois, where he'd lived as Robert Hunter's young indentured servant.

Not on the itinerary was Battle Creek, Michigan, where the Macfadden Sanatorium had finally expired two years earlier. As

1929 began, Macfadden was in talks to procure a much more impressive facility, the massive redbrick edifice known as the Jackson Health Resort in Dansville, New York. It had been built in 1883 by the health reformer James Caleb Jackson, a hero to the young Macfadden. Jackson had been an early American adoptee of hydropathy and the author of alternative health classics such as *The Gluttony Plague; Or, How Persons Kill Themselves Eating* and *How to Treat the Sick Without Medicine*. The building's latest owners, desperate to unload it, had petitioned Macfadden throughout late 1928 and early 1929 to come up and have a look, and when he did, he was smitten. The structure sat amid the green rolling hills of western New York State, on sixty acres overlooking the Genesee Valley just a few miles from the Finger Lakes. Macfadden agreed to assume the property's various mortgages and took possession in May. He installed graduates of his Physical Culture Institute to run the hotel and maintain a physcultopathic regimen: long walks, hydropathy, sunbathing, spine strengthening, fasting, and the milk diet. An approving early visitor noted that the renamed Bernarr Macfadden Hotel employed "attendants specializing in methods for treatments of Anemia, Asthma, Autointoxication, Catarrh, Constipation, Diabetes, Heart Diseases, Influenza, Insomnia, Obesity, Paralysis, Rheumatism, Stomach Diseases, Thinness and all non-contagious diseases." Guests soon began arriving via railroad and automobile. His *Mirror* rival, William Randolph Hearst, had San Simeon, a California dreamland stuffed with priceless antiquities and exotic animals; Macfadden had Dansville, a physcultopathic Eden loaded with dumbbells and raw milk.

Intoxicated by the influence and access to power the *Graphic* had given him—the proprietors of even the most ridiculous big-city news outlets tend to get their calls to Washington, D.C., returned—Macfadden poured much of his seemingly limitless capital into building a newspaper empire. The *New York Daily Investment News*, a financial tabloid, was rushed into production to cash in on the public's obsession with the stock market.

He'd purchased a string of dailies in Michigan, a paper in New Haven, and the *Philadelphia Daily News*. The last of these was a money-losing tabloid that quickly became profitable through cost cutting. Macfadden summoned its chief executive, Lee Ellmaker, to New York to squeeze savings out of the *Graphic*. By the end of 1928, there were signs that the pink paper might start seeing black ink in 1929. An investment prospectus that the company published boasted that "The *New York Evening Graphic* is about to turn the corner."

In March 1929, Macfadden telephoned Gauvreau with a secret mission. He was to have a conversation with William Randolph Hearst's lieutenant, Arthur Kobler. Gauvreau attended a dinner party at Kobler's Park Avenue apartment, and afterward was escorted into his host's study for a private chat. Hearst, he told Gauvreau, would like to buy the *Graphic*. His offer was $2 million. Gauvreau relayed this information to Macfadden. He countered with a price of $4 million. Neither owner was willing to budge, and the matter was dropped. Macfadden's rationale is easy to fathom. Hearst was notorious for overpaying when he really wanted something. And the *Graphic*, once Macfadden got it turned around, was anticipated to be the company's largest source of income for the 1930s, as well as the megaphone through which Macfadden would shout his political message.

The *Graphic* had two invaluable assets, Gauvreau and Winchell. Those assets despised each other more with each passing week. To Gauvreau, Winchell epitomized the gutter journalism that he both loved and loathed. Winchell, who couldn't resist referring to his boss as a "cripple," felt that a raise had been reneged on and that Gauvreau went out of his way to torture him with petty acts of disrespect, such as rejecting expenses. With two years left on his *Graphic* contract, Winchell signed a deal to jump ship to Hearst, then set out to try to break his contract with the *Graphic*. The gossip columnist started phoning Macfadden at three a.m. and screaming at him. The breaking point seems to have been a call in which, according to a version of the story that

Gauvreau enjoyed telling but could never verify, Winchell "accused the physical culturist of having eaten a planked steak floating in Worcestershire sauce while he was supposed to have been on a diet." That morning, supposedly, Macfadden ordered Gauvreau to fire Winchell. In a matter of days, Winchell's column began appearing in the *Mirror*.

One month later, Gauvreau resigned abruptly, and gave interviews in which he expressed great confidence that he'd soon be running another newspaper. To Winchell's horror, Gauvreau took the job as editor of the *Mirror*. Winchell smelled a rat, recalling how surprisingly well informed Gauvreau had seemed to be at their parting about Winchell's future job duties. Macfadden detected the faint whiff of rodent too. "Gauvreau sold me out," he told a reporter years later. "[He] fixed it so that I'd fire Winchell."

Midway through 1929, Macfadden's media empire was at its peak. The combined annual circulations of his magazines and newspapers was 220 million. He employed 2,500 people. His magazines were fat with advertising. *True Story*, *Physical Culture*, and *True Romances* were on track to exceed their record-breaking numbers of 1928. Even the *Graphic* showed its first monthly profit. Macfadden Publications stock was now up 2,280 percent in ten years. The family purchased an enormous new estate in Englewood, New Jersey, just across the Hudson River from the soon-to-be-completed George Washington Bridge. Macfadden knew the area well—he'd invested half a million dollars in property nearby, on the assumption that the bridge would drive already skyrocketing land values even higher. He'd then placed ads in his magazines asking readers to buy shares in his venture.

An extraordinary trio of biographies devoted to the life of Bernarr Macfadden were published in late 1929. Each described its subject in exclusively glowing terms, which was not surprising—he had bankrolled all three. *The True Story of Bernarr Macfadden*, by Fulton Oursler, was an overview of its subject's rags-to-riches life that stressed his status as a self-made man and enemy of the medical establishment. *Bernarr Macfadden: A Study in Success*, by Macfadden's friend Clement Wood (no relation to Susie), aspired to chisel the image of a certain hawk-nosed profile onto the Mount Rushmore of great business leaders, a pantheon whose inhabitants would rightly include (in no particular order) Rockefeller, Ford, Edison, Woolworth, and Macfadden. In his work for physical culture, Wood argued, "Bernarr Macfadden, from his life and aim and achievement, should be properly hailed as the restorer, the rebuilder, the guardian of the living and necessary foundation of all human existence."

Taken together, Wood's and Oursler's books served as the boldest statement of Macfadden's nascent political platform. Candidate Macfadden stood in favor of states' rights, nudity, long walks, strong leadership, and birth control, and against overeating, doctors, vaccination, Prohibition (he thought it drove people to drink more), premarital sex, small talk, and sleeping with the windows shut.

The third book, *Chats with the Macfadden Family*, by Fulton Oursler's wife, Grace Perkins Oursler, also took pains to describe its subject as "a genius" of health and business, but more to build a case for him as America's greatest father. His vigorous children, untainted by vaccines or prudish influences, were said to delight in their homeschooling and wholesome diet. Macfadden was the only aspiring politician in American history who could have uttered the following: "It has happened upon occasions that the children have come into my room, or come upon me on the sun porch when I am naked, and I have made it a point to pay no attention, to show no embarrassment, and to talk to them as casually as I would at the dinner table." *Chats* also provided the only known instance of a man considering a run for public office who circulated a nude photo of himself.

Perhaps the most interesting aspect of the book is its description of Mary, which is so at odds with her own vision of herself that it seems intentionally cruel. Mary claimed that she didn't read the book before publication and that the vacuous quotes attributed to her within its pages never passed from her lips. Judging from the text—the theme of which could be condensed to "here's the little lady, isn't she something?"—there is no reason to doubt her. Mary is described as a happy homemaker who loves nothing better than to gaze lovingly at her brilliant husband, squeeze babies out between her muscular thighs, and roll up her sleeves (should she be wearing any) to bake a loaf of whole-grain bread. The Mary of *Chats* says she became fast friends with Susie Wood from the time she arrived with Bernarr in America, and cherishes their friendship. Her husband's treatment for a child sick with a cold—cut off her food, give her an enema, and make her sweat out whatever toxin is poisoning her blood—is presented as if Mary applauded loudly after each colonic flushing. The editorial from *Physical Culture* about Byron's death is reprinted in full. For a former swim champion who'd twice packed up her daughters and dragged them out to Battle Creek to reorganize the troubled Healthatorium, the caption that appears beneath one portrait reads like a deliberate slap in the face: "Mary Williamson Macfadden—who asks no career or fame further than being the best wife, mother and homemaker she knows how!"

"**At times, in** its working out, I am inclined to doubt the wisdom of a democracy which emphasizes everything except body-building and health," Macfadden told Clement Wood for *Study in Success*. "There are times when I believe that America needs a Mussolini, as never before." Macfadden, of course, had been a sucker for strongmen since the first time he'd seen George Baptiste hoist that eighty-pound dumbbell back in St. Louis. No man, though, impressed him more than the ascendant Italian dictator Benito Mussolini. As early as March 1927, *Physical Culture* was publishing gushing profiles of the barrel-chested martinet who brought order to an unruly nation and still found time for daily strenuous exercise. "Fascism," he told *Physical Culture*'s correspondent, "is a muscular creed. We believe in discipline. No discipline is possible without complete intellectual, moral and muscular coordination." A follow-up article breathlessly described the thrill of watching thousands of Italian girl scouts give "Muscleini" the outstretched arm salute before putting on a gymnastics demonstration. "Mussolini is not merely the Dictator," wrote the correspondent, "he is the Macfadden of Italy."

Macfadden's donation to the Republican Party paid off with a minor functionary job as a member of President Hoover's Committee on Child Health and Welfare. In September 1930, while touring the European continent in this capacity, Macfadden secured a meeting with Il Duce.

Macfadden's other meetings in Rome, with Pope Pius XI and Italian King Victor Emmanuel III, were cordial, and Macfadden presented the pope with an "absurdly large" photo of his physical-culture family, according to Mary. (She'd been allowed to accompany him on the journey on the condition that she lose thirty-five pounds.) He was much more excited to meet Renato Ricci, Mussolini's undersecretary of state for physical education. Ricci gave the Macfaddens a grand tour of the gymnasiums and camps where Italy's armed forces and youths were trained to respect their bodies.

The meeting with Mussolini, on the evening of September 19 at the Venezia Palace, started awkwardly. Mussolini came out from behind his desk to shake Macfadden's hand, express admiration for his work, and underscore their shared passion for the Roman ideal of *mens sana in corpore sano*. Macfadden apparently was nervous standing in the presence of one of his demigods. "Your soldiers eat too much," he blurted out by way of greeting. Then he offered to invite a large group of cadets to his facilities in the United States to show off what his own physical-culture methods could achieve. Mussolini accepted.

The following February, forty Italian cadets in full military dress from the Fascist University of Rome arrived in New York. Their first evening was spent watching championship wrestling with Macfadden at Madison Square Garden, and was followed by a trip to D.C. to shake hands with President Hoover. From there, they were dispatched to the sanitarium at Dansville and then to Castle Heights. For two months, as Italians back home struggled with the food shortages caused by the deepening worldwide Depression, the black-shirted cadets were deprived of pasta and wine and fed cracked wheat. They went on long walks, lifted weights, and learned to play baseball. They visited the Grand Ole Opry. Their chests expanded, and their waists shrank. The *Graphic* covered the physical-cultural exchange with a two-page story, headlined MUSSOLINI AND MACFADDEN TRY A NOBLE EXPERIMENT TO PREVENT WAR. The next year, Mussolini instructed the king to bestow upon Macfadden the decoration Commander of the Order of the Crown of Italy.

Chapter 13

New Deals

No One Sends Flowers to a Fat Girl . . .

—TYPICAL AD IN *PHYSICAL CULTURE* DURING THE MID-1930S

While Mussolini's cadets were slimming down, Bernarr Macfadden focused on strategic growth. He had his eye on
Liberty, a general-interest weekly magazine that had been started
in 1924 by newspaper barons Colonel Robert McCormick of the
Chicago Tribune and Joseph Patterson of the *New York Daily
News*, who were cousins. Their intent was to create a competitor
to the spectacularly successful *Saturday Evening Post* and its rival
Collier's, magazines that dominated readership in the era just before the newsweeklies exploded. In selling 2.5 million magazines
each week, *Liberty* had built the largest single-copy sales of any
publication in the country. (Single-copy sales are considered more
important than subscriptions in the magazine business because
each purchase is an active choice to buy and read an issue, as opposed to the more passive experience of subscribing.) *Liberty* had
never come close to breaking even, however, since its price was a
nickel—compared to twenty-five cents for *True Story*—and advertisers found its competitors more attractive.

Macfadden traded his *Detroit Daily Illustrated* tabloid to Mc-
Cormick and Patterson for their magazine. Then he called Oursler

to let him know about the swap. "Charley, I've bought *Liberty*," he told him. "You're the new editor in chief. But you'll keep on with the other magazines, too. We couldn't afford to lose you on the others."

As supervising editor of Macfadden Publications, Oursler was already overseeing nine titles. He was also plotting a new novel and a new play, writing detective fiction under the nom de plume Anthony Abbott, and churning out his usual quota of magazine articles. Much of this he accomplished without walking out his front door. He'd purchased a new home, a brownstone building just a few blocks away from the Sixty-fourth Street headquarters, at 217 West Seventieth Street in Manhattan. All day long, messengers would arrive with stacks of manuscripts to be processed, which a team of secretaries would keep moving through the editorial pipeline and hand off to other messengers. Visitors from the political and literary worlds would pop in for conferences with Oursler throughout the day and night. And yet when *Liberty* was added to Oursler's workload, it seemed to energize him even more. "He kept all the balls in the air at the same time, and he did it with a riotous delight in his own dexterity," wrote his eldest son, Will. "He lived as if he had a private wire to some cosmic powerhouse."

The first issue of *Liberty* as a Macfadden publication was dated April 4, 1931. Within a month, Bernarr Macfadden began publishing a one-page editorial each week, covering topics that ranged from the evils of organized crime to the importance of getting Americans back to the land to grow their own food. Nine out of ten had a political spin. "Through my editorial page," he later bragged to his friend, the future New York governor and presidential candidate Thomas Dewey, "I believe that I have more influence politically on the masses than any one individual in the United States."

'Oursler's imprint on *Liberty* was visible almost immediately. Patterson and McCormick's magazine had been a stodgy, cheaper version of the *Saturday Evening Post*. Oursler's genius was in repackaging *Liberty*'s contents. He knew how to twist a well-worn current event to the precise angle that made the story sound fresh. The old *Liberty* might have published a long hand-wringing story about how law enforcement was battling the gangster menace. Oursler assigned an interview with Chicago crime boss Al Capone on how he'd govern America if given the chance. (Capone said he'd get rid of corrupt politicians, for starters.) Because of the logistics of printing and shipping millions of copies, *Liberty*'s pages were sent to the printer several weeks before they appeared on newsstands. The result in its first years was a moldy magazine that varied little, aside from the changing seasons depicted on its too-cute covers.

Oursler had a knack for smelling where the news was headed. He published an incendiary interview with a rising young European political star named Adolf Hitler, titled "When I Take Charge of Germany"—though the fact that the interview had actually taken place in 1923 was conveniently omitted. In early 1932, General Billy Mitchell wrote a pair of articles forecasting an air attack by the Japanese. International figures were signed up as regular contributors: Winston Churchill, H. G. Wells, and, somewhat later, Mahatma Gandhi filed regular dispatches. Gandhi wrote one of *Liberty*'s most famous stories, a treatise on celibacy that Oursler splashed on the cover as "My Sex Life—by Mahatma Gandhi." Under orders from Macfadden, physical culture often scored a point or two in an unlikely place. "People like Hitler, and Stalin, Mussolini, Chiang Kai-Shek, and Hirohito would smile and talk a little when asked what they ate for breakfast and what they did to keep in condition under the great strain of their important work," recalled Cornelius Vanderbilt, the magazine's globe-trotting correspondent.

The public figure with whom the magazine had the most mutually profitable relationship was the governor of New York, Franklin Delano Roosevelt. In the summer of 1931, Roosevelt, one of the most admired Democratic politicians in the United States, was among the front-runners to earn the nomination to run against the mortally unpopular Republican incumbent, Hoover. Lingering over FDR's prospects, though, was one question, addressed in an irresistible headline in *Liberty*'s July 25, 1931, issue: "Is Franklin Roosevelt Physically Fit to Be President?" Political rivals had started a whisper campaign that wondered whether a man stricken with polio was up to the task of running the country. (*Time*, reliably conservative, had already chimed in with the opinion that he was "utterly unfit.") In light of the lengths to which the Roosevelt team went in later years to obscure the president's paralysis, the frankness with which FDR spoke to *Liberty* is striking. He discussed his leg braces and half-jokingly noted how not being able to stand up and leave one's desk had certain productivity benefits. "If the paralysis couldn't kill him, the presidency won't," his wife, Eleanor, not known for her jocularity, told *Liberty*'s writer. Oursler arranged for a team of doctors to examine FDR. They found him in perfect health. The biggest doubt about the Roosevelt campaign vanished almost overnight.

Roosevelt understood immediately the power that *Liberty* possessed to help his career, and began cultivating its owner and editor. Years later, the president admitted to Oursler that he had figured the *Saturday Evening Post* and *Collier's* would support the candidate preferred by their snobbish advertisers—Hoover. So he laid his bet on the third horse in the race. The flirtation with both Oursler and Macfadden continued throughout 1931; the latter was invited to Roosevelt's hydrotherapeutic waters resort at Warm Springs, Georgia, for a visit in October. When *Liberty* prepared articles about both Hoover and Roosevelt for an election preview, Oursler sent prepublication manuscripts of both profiles to each party for comments. The Hoover White House returned its copies in an envelope, unread, with no note. Roosevelt invited Oursler over to his Manhattan home to discuss the stories in his study. Not long after, Roosevelt's close associate Louis Howe schemed to have Macfadden—a committed Hoover man—seated next to the governor on the platform at a Red Cross event. "I think Roosevelt is the coming man," Macfadden told Oursler afterward. "He and I see eye to eye on everything. He is an ardent devotee of physical culture and always has been." Roosevelt's bet on the long shot paid off. He signed a contract to write essays on political topics for *Liberty*, which during 1932 published seventeen of his articles in its pages. Many of them, with titles such as "Banishing the Dole," promised that FDR would be a tax-slashing friend of big business, which delighted Macfadden.

Shortly after the Red Cross event, Macfadden asked Oursler to meet with Eleanor Roosevelt. Over lunch, Oursler learned that his boss's own mysterious powers of persuasion had been at work. Macfadden had suggested to Mrs. Roosevelt that she become the editor of his new, "decidedly high class" child-care magazine, called *Babies, Just Babies*. Mrs. Roosevelt said she was interested, and FDR approved. Oursler negotiated the terms of her contract with Louis Howe. She was to receive $500 per issue, but would get a bump to "$1,000 per issue if you are at that time of such renewal living in the White House." The contract also stipulated that her daughter, Anna Roosevelt Dall, would be hired as her secretary at $200 per month.

The first issue of *Babies, Just Babies* appeared in late September, a little more than a month before the 1932 presidential election. The magazine delivered on the promise of its name. "Babies! Can you think of anything more wonderful?" gushed Mrs. Roosevelt in her commencement editorial. She told a story about how Anna, as an infant, had disrupted a dinner party with her crying, and how Eleanor had soothed her by turning her over a knee and patting her back. Articles included advice ("A Layette for $11.10? Here's How") and, naturally, a true-life tale ("I Became a Mother at 42"). A dozen photos of adorable babies in various states of undress were printed. Mrs. Roosevelt read and approved every page before it was published.

Just Babies, unfortunately, was a source of more ridicule than revenue—the *Harvard Lampoon* published a parody called *Tutors, Just Tutors*—and while it did nothing to halt her husband's landslide victory, Eleanor Roosevelt clearly didn't enjoy the mockery. In early 1933, after only a few issues, she wrote to Macfadden that withdrawing from the project seemed "the most sensible thing to do." Macfadden accepted her resignation, and *Just Babies* came to an early end.

The economic downturn that followed the stock-market crash of October 1929 had seemed reversible throughout most of 1930. By 1931, however, a bad recession had hardened into the Great Depression. Wall Street's tailspin had whittled down Macfadden's net worth and ended his spending spree, though Macfadden Publications was still the third-largest magazine publisher in the United States. The company now owned nineteen titles—and its assets also included six newspapers, the Dansville home, the Castle Heights Academy, and various small physcultopathic training schools.

In the autumn of 1931, Macfadden used an appearance on WOR radio to make a surprise announcement. He was giving away $5 million to fund the new Bernarr Macfadden Foundation, a nonprofit organization devoted to "the perpetuation of physical culture and health building." The grant would be made in the form of Macfadden Publications stock and his personal real-estate holdings. His various schools, camps, restaurants, and educational facilities would be run by the foundation. Scholarships would be granted to aspiring physcultopathists, without "discrimination as to age, race, nationality, sex, creed or color of the beneficiaries." An additional $1.5 million in securities was to be set aside for his dependents, though these would be recalled by the foundation in the unlikely instance of a financial crisis.

The $6.5 million evidently accounted for almost all of Macfadden's earthly holdings. After making these two bequests, he said, he expected to live on a salary for the rest of his life. "It is a source of indescribable relief to feel like a free man again," he told a reporter. "Too much money unwisely used makes people greedy and ungrateful, destroys the home, steals your happiness, enslaves, enthralls you, lowers your vitality, enfeebles your will, leaving you ultimately but little more than a flabby mollycoddle."

Mary Macfadden did not share her husband's sense of inner peace. The two had been fighting more frequently in the years after Byron's death, and after the Mussolini trip—from which Mary had returned home alone following a spat in Paris—a decade of simmering resentment reached the boiling point with the announcement of the foundation. Macfadden saw the foundation as the culmination of the partnership he and Mary had forged during their walks in the English countryside back in 1912. Now that their physical-culture family was complete and physcultopathy had been perpetuated for decades beyond its founder's life span, they had little use for millions.

From Mary's perspective, her husband had just given away everything they'd worked for. The bequest had included a large block of stock that she'd transferred to him to help cover losses as the value of his New Jersey real-estate holdings cratered. (The state eventually auctioned off his land when Macfadden refused to pay a $526,000 tax assessment.) With profits down at *True Story*, the *Graphic* once again burning cash, and *Liberty* needing regular subsidies, he had signed away their comfort in the name of physical fitness.

The conflict escalated rapidly. The Macfaddens moved into separate quarters in the Englewood mansion. Bernarr had hired a live-in trainer to accompany him while walking the eleven miles from Englewood to his office each day. Mary insulted the coach and kicked him out of the house. In one of Bernarr's capital-letter editorials not long before their split, he wrote: "ONE OF THE FIRST DUTIES OF A WIFE IS TO MAINTAIN THE SEX APPEAL WHICH SHE HAD DURING HER SWEETHEART DAYS. . . . Marital happiness depends upon a full recognition of this imperative requirement." Mary was having trouble controlling her weight due to a weakness for sweets.

She frequently insisted to dinner guests that *True Story*, the source of most of their wealth, had been her idea, which Bernarr said he found "humiliating." She was also convinced that Oursler, whom she now despised beyond any rationality, had hypnotized her husband into believing that he was a man of political destiny. And Mary blamed Oursler's mesmerism for the shaky investments in the *Graphic* and *Liberty*—and the rift in her marriage.

Above all, the Macfaddens couldn't agree on how to raise the children. Materially, even after the foundation was created, the family lacked nothing. The new twenty-eight-room home in Englewood sat on eighteen acres, featured a solarium, tennis courts, a miniature-golf course and putting green, a mirrored dance studio, a private lake, waterfalls, a fifteen-foot-deep swimming pool, rose gardens, an acre-size vegetable patch for Macfadden's ration of pesticide-free roughage, and a four-car garage, above which sat an eight-room apartment. But Mary had grown to see her marriage as a physical-culture dictatorship in which the children's needs were always secondary to their father's demands.

A man who'd spent his own lonely childhood working and discovering the glories of physical culture assumed that his children would match his success if they followed a similar, if more refined, path. "My father was always putting us on different diets, the milk diet, raw food diet, meat diet and fasting, which he said would purify your blood," remembered his daughter Byrnece years later, with no great fondness. "We had dancing lessons, swimming lessons, tennis lessons and went for long walks all the time." Macfadden demanded showpiece offspring—he wanted "cultivated vocalists," "good pianists," and "expert dancers" who were "athletic in every way"—yet Mary refused to enter the children in athletic competitions. She argued that he only "wanted to use his family as guinea pigs" to garner publicity. The daughters found their father a tyrant. With him out of the house so frequently, however, according to one of Bernarr's spies, Mary couldn't control them. After living completely sheltered lives into their teens, they were on the verge of womanhood.

Macfadden was an enthusiastic employer of private detectives and collectors of intelligence—he secretly kept a sleuth on staff at his office and had his advertising salesmen tailed—and the agent he maintained in the Englewood house filed regular dispatches on his daughters' behavior. Braunda, aged fifteen, though her father's favorite, was a spoiled spendthrift. Beverly was "pretty much a hoodlum," the spy reported, a thirteen-year-old whose "principal object in life seems to be to get somebody to spend large amounts of money on things suitable only for a chorus girl of nineteen with a rich sugar daddy on a string." Byrnece, the eldest of Macfadden's children with Mary, chafed under her father's rule, an attitude perhaps not shocking in a teenage girl greeted by a life-sized statue of her naked pubescent self each time she sat down to a meal. Her father thought that, at age seventeen, Byrnece was just about ready to start socializing regularly with people her own age. Instead, she got engaged to Louis Muckerman, an investment broker. Hearing the news, her father wrote Byrnece that his "only desire is for your future happiness," then explained that she was nowhere near in good enough physical shape to take on the duties of wife and prospective mother. "To meet with my approval you should have at least three months vigorous training. Six months would be better. Walk five miles daily, gradually increase to ten."

During one fight in Englewood, Mary hurled a box of razor blades at Bernarr, breaking two of his teeth. She said she'd only done so to defend herself after he'd struck her and waved a chair menacingly. On another occasion, Mary pulled a gun from a drawer. Bernarr wrestled it away; she claimed she'd wanted to kill herself. Bernarr told one of their daughters that he was certain that Mary was trying to kill *him*. He moved out. In December, *Time* reported that Mary had suffered a "nervous collapse" in a Manhattan theater.

In April 1932, the father of the perfect physical-culture family swallowed his pride and agreed to a separation.

The *Graphic* had shown a profit for precisely one month, in 1929. Almost on a whim, Macfadden had appointed twenty-eight-year-old mortgage-bond salesman William Robinson as president of the paper. Robinson engineered an impressive turn-around of the advertising department. For the first time, cigarettes appeared in a Macfadden publication, in a nonexculpatory way: Chesterfield, Old Gold, and Lucky Strike all shared space with L'il Samson puffing out his pectorals. When Macfadden asked Robinson to report to Gauvreau's penny-pinching former nemesis from Philadelphia, Lee Ellmaker, though, Robinson resigned. Macfadden's loss was Robinson's gain; he went on to run the *New York Herald Tribune* on his way to becoming CEO of Coca-Cola.

As the postcrash economy tumbled toward Depression, newspaper advertising dried up like Oklahoma topsoil. By early 1932, the anything-goes sentiment of the twenties was as dead as the corset. Nonstop coverage of Lucky Lindbergh's aerial triumph had long since given way to nonstop coverage of the Lindbergh baby kidnapping. Wisecracking, hard-living mayor Jimmy Walker and his laissez-faire city government were no longer funny; Walker was now the subject of the very serious Seabury corruption investigation, a man whose former pals at the *Graphic* now printed banner headlines like $300,000 GIFTS WORRY WALKER. The Dow Jones Industrial Average bottomed out at 41.22 from its 1929 high of 381.7. Average weekly wages had slipped nearly 40 percent at the same time. The thousands who were crowding into Macfadden's penny restaurants for a bowl of oatmeal with raisins and cereal coffee no longer had the luxury of plunking down two cents to snicker at *Graphic* stories such as TWO WOMEN IN FIGHT, ONE STRIPPED, OTHER EATS BAD CHECK. Circulation, which had stabilized at 350,000 as the Depression started, began to fall. So, for the first time, had the profits of Macfadden Publications.

At the start of 1932, Macfadden announced in a series of ads in the *Graphic* that after spending most of 1930 and 1931 away from the paper—"I have not been in the *Graphic* office half a dozen times in the last two years"—he was rolling up his sleeves and taking charge in the New Year. To celebrate he began a "lucky bucks" contest in which he would give out $100 prizes to readers. For a few months, the *Graphic* cleaned up its act to the point where it looked like a vigorous pink sibling of the *Daily News*. Advertisers responded positively. One evening in June, Macfadden convened the 420 employees in the paper's auditorium and announced a plan under which they would become owners of the *Graphic*. Most had already taken pay cuts; in lieu of further trims, they'd henceforth contribute 10 percent of their salaries toward buying stock in the paper. Under this program, the *Graphic* would be completely employee-owned by 1945.

Instead, on July 7, the *Graphic* abruptly ceased publication. The decision was evidently not planned far in advance, as Macfadden printed no valedictory editorial thumbing his nose at the medical trust (and certainly nothing up to the standards of the prior month's spirited AMELIA EARHART'S TOMATO JUICE DIET); the last bit of prose on the last page of the *Graphic* was a boilerplate diatribe from the owner against overtaxing businessmen. The red ink from the *Graphic* would provide write-offs for years to come. Macfadden estimated that he'd lost between $7 million and $8 million on the paper. His comptroller, factoring in the cost of printing plants and real estate, calculated the figure at closer to $11 million.

Chapter 14

Empire Builder

In 30 years Bernarr Macfadden's taste in presentation has improved. Instead of photographs of the publisher clad only in an abdominal supporter to illustrate his lectures on "physcultopathy" there are chastely presented "personal messages" [in Physical Culture]. . . . Instead of testimonials of persons who "cured themselves" of asthma, rheumatism and appendicitis by "natural methods," there are articles on dietetics, child guidance, prevention of tooth infection, by qualified authorities.

—*Time*, September 21, 1931

In the dark final hours of Herbert Hoover's presidency in March 1933, the United States economy wobbled as if preparing to expire. Thousands of banks had failed, and governors across the nation ordered their states' remaining financial institutions shuttered, to prevent the virus of runs by depositors from spreading. The New York Stock Exchange closed its trading floor; no one could say when it might reopen. Unemployment reached 25 percent.

Amid the gloom, the Macfadden Foundation's restaurants were feeding ten thousand persons a day at a few cents a head, leading Bernarr Macfadden to announce that with a little federal assistance, he could provide meals to a million hungry Americans every day. "I would sign a contract on that basis as long as prices remain as they are today if the government would furnish the buildings and equipment," Macfadden told the *New York Times*. Within months, without government aid, he expanded operations to eight restaurants in three cities, including one in Washington, D.C., that catered exclusively to a black clientele. The subsidized eateries remained open until the Depression eased and customers, having had their fill of oatmeal and prune juice, stopped coming.

Like many business leaders, Macfadden believed Roosevelt's campaign vow to jump-start the economy with a heavy dose of budgetary discipline. "The new deal that our President has promised . . . will mean, first and foremost, that federal government expenses will be 'cut to the bone,'" Macfadden wrote in a February *Liberty* editorial. A few weeks later, in a postinaugural valentine titled "The New Deal Is Here—God Be Praised!" Macfadden exclaimed that "the dictatorial powers placed in [Roosevelt's] hands have been used forcefully and intelligently. . . . Our new leader has already proved himself magnificently—bold, courageous, sagacious, and marvelously capable." In mid-May Macfadden was delighted when Eleanor Roosevelt responded positively to his loud hints that he'd appreciate an invitation to the White House. Berwyn and Brewster splashed in the presidential pool during the weekend FDR was finalizing legislation for his Tennessee Valley Authority bill.

About the only courtesy Macfadden refused to extend Roosevelt in those first hundred days was to lend him the services of Fulton Oursler. Roosevelt offered Oursler his pick of ambassadorships, and Oursler selected Siam. When the White House ran the idea past Macfadden, though, the publisher vetoed it.

Oursler, who seemed to be allergic to working at the Sixty-fourth Street offices, found another method of staying far away. The previous summer, RKO Pictures' boy-wonder chief, David O. Selznick, then working on *King Kong*, had inquired if Oursler would be interested in coming out to Hollywood to adapt his novel *The Great Jasper* for the screen. Oursler accepted, Macfadden evidently agreed, and the two men saw no reason that Oursler couldn't continue overseeing the stable of Macfadden magazines from three thousand miles away. For six months, in what amounted to his spare time between writing assignments, Oursler edited *Liberty* and supervised almost a dozen other magazines via letter and telegram. Oursler returned east in March 1933, but his stay in New York City was brief. He had purchased a modest home in West Falmouth, Massachusetts, which he christened Sandalwood and proceeded to reconstruct into a thirty-room monster. Oursler installed a teletype machine, and for the next seven years, editors in Manhattan would gather around the printer on their end every day at 10:00 a.m. and 4:00 p.m. as page after page of literary marching orders clack-clacked out of the ghostly ether. Sandalwood soon graduated to a Macfadden Publications satellite office, and many a Cape Cod resident did a double take at the sight of a muscular older gentleman with wild white hair, jogging or racewalking barefoot on the road from the airstrip at Woods Hole, tailing the chauffeured car that had been sent to fetch him.

With Macfadden keeping a tight rein on costs, *Liberty* eked out profits for most of 1932. The rest of his company was doing less well. Consumer spending was plummeting along with the economy, and so, for the first time, were *True Story*'s sales. By April of 1933, preferred shares of Macfadden Publications stock that four years earlier had sold for $57 were now being offered at $12. The company had also defaulted on paying its annual $6 per share dividend.

On May 9, the creaky détente between Bernarr and Mary collapsed. Mary filed suit in Trenton, New Jersey, demanding a financial accounting under the terms of their previous year's separation agreement. Mary was supposed to be receiving $15,000 a year, based on a trust fund of 2,500 shares of Macfadden Publications preferred stock. Because of the missed dividend, however, only $10,330 had been disbursed over the first twelve months. In her filing, Mary accused her husband of manipulating the stock price down to minimize her dividend payments.

Macfadden's response was typically vigorous. He sued for divorce on grounds of infidelity, and cited an obscure Swedish nobleman identified in his complaint as "Baron Rosencranz" as his wife's alleged lover. Mary countersued, claiming that her husband had been unfaithful with an employee of the Physical Culture Hotel at Dansville, whom she'd seen sneaking out of her husband's room late at night. Macfadden counter-countercharged that Mary had also committed adultery with a West Point cadet. Even worse, he added in his court filing, Mary had undone all his hard work to make the children national paragons of fitness by discouraging their physcultopathic regimens and allowing them to "smoke and drink in swanky speakeasies."

Mary's alleged infidelities were likely fictions, a legal thrust by her husband intended to draw first blood. Though he almost certainly had his wife under surveillance, the first man Macfadden hastily named as a paramour was, in fact, dead. (He even spelled his supposed rival's name wrong—it was Rosencrantz, with a *t*.) Late in life, when Macfadden had married for the fourth and last time, he would accuse *that* wife of leaving her bedroom window open to facilitate visits from a lover, although they lived on the fifteenth floor.

Macfadden's infidelities, on the other hand, were simply assumed by his friends and employees. His Physical Culture City romance with Susie Wood was evidently an early chapter in a long history of philandering. Cornelius Vanderbilt wrote of rumors of Macfadden's fling with a "Russian princess." Orr J. Elder's son recalled that it was "impossible for my father to keep any young girls around with Macfadden." Even older ladies near him were suspect. The crabby white-haired manager of the photo studio in the basement of the Macfadden Publications Building was whispered to be one of the boss's former flames. His fourth wife intercepted love letters and caught him in bed naked with another woman.

Macfadden at age four; as an
adolescent in St. Louis with
his uncle Crume Miller; at age
eighteen; and photographed by the
great Sarony at age twenty-five (the
lion was painted in afterward).

Three great influences: Sylvester Graham, the Johnny Appleseed of American vegetarianism; strongman Eugen Sandow, whose stage show at the Chicago World's Fair redirected Macfadden's life path; and Anthony Comstock, Macfadden's prudish nemesis, seen in a cartoon from the original *Life* magazine.

St. Peter: NO, ANTHONY, NO. WE MAY HAVE THINGS HERE YOU WOULD OBJECT TO.

WEAKNESS A CRIME
PHYSICAL CULTURE 5¢

PHYSICAL CULTURE PUBLISHING CO., Townsend Bldg., 25th STREET AND BROADWAY, NEW YORK CITY, U.S.A.

OUR PREMIUM
Physical Culture Watch

This is a finely polished black gun-metal watch. Not a toy. Tells you what time to rise, to eat, to exercise, etc. Warranted for one year. Has the appearance of and keeps time as well as a $10 or $15 watch. Attractive and useful.

Photograph of watch. Exact size.

We have purchased a large quantity of these watches at a low price, for the purpose of offering an attractive and useful premium for subscriptions.

Though hour hand is not shown in cut it is part of the watch.

This watch sent postpaid upon receipt of two yearly subscriptions or for one yearly subscription and 50 cents additional.

PHYSICAL CULTURE PUB. CO.,
DEPT. F, PHYSICAL CULTURE CITY, SPOTSWOOD P. O., N. J.

A typical copy of *Physical Culture*, ca. 1900; one of the many healthy premiums offered to its subscribers (note the very specific instructions on the watch face); an advertisement for the utopian Physical Culture City.

Mary Williamson Macfadden, shortly after being crowned Great Britain's Perfect Woman by her new husband; a 1912 advertisement for *Macfadden's Encyclopedia of Physical Culture*; a Healthatorium patient, starved down seventy-two pounds in six weeks on a water-only fast.

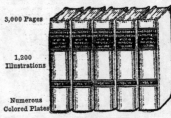
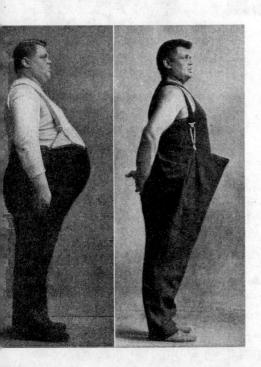

Macfadden with Fulton Oursler, his editorial genius and the soon-to-be enemy of Mary; father and daughter statues, commissioned by Macfadden to show off the perfect physical development of himself and his twelve-year-old daughter, Byrnece; the roster of Macfadden Publications in the late 1920s.

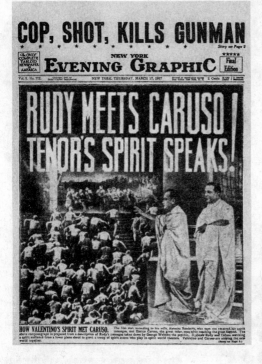

The *New York Evening Graphic*'s exclusive report on Rudolph Valentino's ascension into heaven; *Graphic* editor Emile Gauvreau; a classic composograph, starring Peaches and Daddy Browning and their opinionated pet gander.

The Macfadden clan, around the time its paterfamilias caught the political bug; one of eight Macfadden Foundation restaurants, which served cheap, healthy meals through the darkest days of the Depression; a squad of extremely fit Italian cadets offers a cheery Fascist salute to their host, at center, before departing back to Rome on the SS *Biancamano*.

These photographs, taken just before going to press, show clearly the vigor and superb condition of the noted publisher, as he is today at 70

PHOTOGRAPHS BY MACFADDEN STUDIO

Macfadden in triptych, flexing in *Physical Culture* to mark his seventieth birthday; with fourth wife, Johnnie Lee Macfadden, prior to his first parachute jump at Dansville; near the end—his fortune gone, his fame fading, and his beautiful physique slipping away.

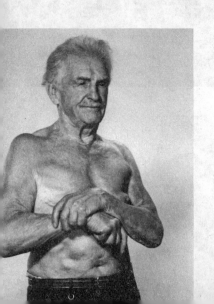

Macfadden's suit against Mary was certainly not buttressed when he was sued the following year for alienation of affections by Soitir Adams, a Boston cafeteria manager. Adams's wife, Abbie, had been employed in the 1920s as a nurse at what remained of Macfadden's Battle Creek Sanitarium. It was there, Mr. Adams claimed, that Macfadden had seduced his wife. He wanted $100,000 for his suffering. Mrs. Adams seemed to feel differently, according to one newspaper account, which reported that in her deposition she "said she had left Adams and had been intimate with Macfadden because she liked him better." *Time*, ever eager to link the subjects of Macfadden and sex, mentioned that she admired his "technique."

Macfadden eventually won the Adams suit, but he folded against Mary. He paid the money he owed her, and their strange nonmarriage continued. *Why* Mary insisted that they remain legally attached is hard to fathom. She seems to have worried that if Macfadden died—his impressive musculature notwithstanding, he *was* twenty-five years older—her alimony payments would cease. In the pages of Macfadden magazines, once-ubiquitous references to their dream marriage had vanished; Macfadden's memoir "My Fifty Years of Physical Culture," which was serialized in *Physical Culture* during most of 1933 and 1934, doesn't mention Mary's name once.

Whereas Macfadden's first serialized life story in *Physical Culture*, which ran in 1914, traced his path toward the enlightenment of physcultopathy, this second surge of recollections consisted largely of Macfadden talking about his new passion for flying. After a few years winging around the country as a passenger and assembling an impressive collection of aircraft, he'd obtained his pilot's license in 1931, at age sixty-two. His zest for air travel always outweighed his skill at the controls. He frequently leaned his head far over the side, searching for rivers and railroad tracks by which to navigate. "He was always tilting the plane so he could look at the ground for landmarks," said one executive who reluctantly flew with the boss. Another employee recalled seeing a colleague drop to her knees and pray during a bumpy hop from New York City to Montauk, at the end of Long Island. Macfadden's takeoffs and landings weren't exactly textbook quality, either. By the end of the thirties, he'd suffered five crack-ups at airports around the country—hitting a tree, plowing through a pair of fences, and flipping at least two craft. (Neither of which was the Northrop Gamma that Howard Hughes had used to set a cross-country flying record, which Macfadden purchased in 1938 to compete in long-distance races.) Miraculously, no passengers ever suffered more than cuts and bruises.

By the mid-1930s, *Physical Culture* had evolved once again, this time from the shrill, antidoctor soapbox of the late 1920s into a sophisticated consumer product focusing on pure food, preventive medicine, and other topics appealing to women as well as men. (The November 1931 issue contained a long story on the mysterious Eastern discipline of yoga, including illustrations of a flexible female demonstrating how to execute a perfect *sarvangasana*, or shoulderstand.) Oursler's deft editorial hand was visible in the cover art, which tended toward buxom pinup lasses merrily vaulting pommel horses or drawing back archery bows—often dropping a shoulder strap or flashing a bit of cleavage—and in irresistible feature stories such as "Can a Wrestler Beat a Boxer?" and "I Bought a Baby for Christmas." (Ample room remained for the usual panoply of Macfadden's trademark oddities, such as "America, Too, Goes Nudist" and "Physical Culture Banished My Goiter!") *Physical Culture* sold an average of 340,000 copies per month in 1933, an astonishing number for a magazine devoted to alternative health ideas. By comparison, *Hygeia*, with the might of the AMA behind it, never topped 90,000. A *New Yorker* cartoon from July of that year reflects *Physical Culture*'s place in popular culture: As two muscular fellows square off for a fistfight, a newsboy sensing an easy sale approaches with magazines in each hand. "*Physical Culture*, gentlemen?" he asks. After

John Harvey Kellogg's famed Battle Creek Sanitarium went into receivership in February 1933, Macfadden could lay uncontested claim to being the most important alternative health figure in America.

His influence extended well beyond the pages of his magazine. The Physical Culture Institute, a state-of-the-art culinary laboratory in a Manhattan high-rise (dressed up as the sort of suburban kitchen in which the cupboards held blackstrap molasses and wheat germ), was answering thousands of queries about healthy foods, tinkering with meatless recipes, and testing products. More than one hundred Physical Culture Clubs organized through the magazine met regularly in the United States. The Bernarr Macfadden Foundation sponsored a program in which almost twenty thousand Alabama boys exercised regularly on unused airfields. Macfadden operated physical-culture camps near New York City each summer and later opened elementary schools, named after himself and modeled on Castle Heights, in Tarrytown and Briarcliff Manor, north of the city.

Macfadden took a special, if unorthodox, interest in the training of children. When the Portuguese government approached Macfadden with an offer to replicate his Italian cadet experiment on a group of fifty undernourished boys aged ten to fifteen, Macfadden remodeled a ramshackle fort at São João do Estoril, twenty miles outside of Lisbon, into the Macfadden Children's Colony. Pleased with the results he obtained there, he repeated the regimen in Hackensack, New Jersey. About twenty Garden State ragamuffins were housed for several months at the clubhouse of a golf course Macfadden owned. They were stuffed with Graham bread and vegetables, exercised twice daily, and put to bed for at least eleven hours each night. In all cases, the sickly youths gained weight and vitality.

In many ways, the world seemed to be coming around to Macfadden's way of thinking. Arthur Kallett and Frederick Schlink's book *100,000,000 Guinea Pigs*, a warning about the impurities in pharmaceuticals and processed foods, was one of the best-selling books of 1933 and 1934. Physcultopathy had endured long enough to trickle down to a new generation. In a replay of Macfadden's experience at the Missouri Gymnasium, a skinny sugarholic teenager named Jack LaLanne attended an Oakland lecture on nutrition and exercise given by one of Macfadden's most loyal disciples, *Physical Culture* staffer Paul Bragg, and exited with a new sense of purpose. Within a few years he exercised himself into a physique champion, on his way to revolutionizing the fitness industry. Charles Atlas graduated from midsize advertisements in the back of *Physical Culture* to large ones near the front; the strongman auditioned several humiliating cartoon scenarios before selecting his immortal come-on, "THE INSULT THAT MADE A MAN OUT OF MAC." The March 1933 issue of *Physical Culture* carried a small ad in the back for a new magazine published out of York, Pennsylvania. Called *Strength and Health* and influenced by both *Physical Culture* and Macfadden's short-lived publication *Muscle Builder, Strength and Health* and its charismatic editor, Bob Hoffman, would help transform American weightlifting and bodybuilding from sideshows into respectable competitive fields. In a 1935 *Chicago Tribune* story about the retirement of Chicago Bears guard Joe Kopcha, the 220-pound All-Pro lineman told the paper how Macfadden's regimen had helped him to become a star prep athlete. "He went without meat for three years," the *Tribune* reporter wrote, "and when the high school coach objected, Joe curtly told him that if it was good enough for Macfadden it was good enough for him."

The Hotel at Dansville, the primary focus of the Macfadden Foundation, had blossomed into a thriving center of what one advertisement billed as its four Rs: restoration, rebuilding, rest, and recreation. (A fifth R, rehab, could have been added, since intemperate New Yorkers often came up to dry out.) Checking in at Dansville was like boarding the good cruise ship *Physcultopathy*. Visitors typically arrived by train and took a taxi to the hotel, where they were greeted by a large statue of their host clad only in sandals and a wisp of strategically placed cloth. Most guests were scheduled hour by hour for the length of their stays. Once examined by a licensed MD schooled in what promotional brochures called "natural methods," guests were encouraged to visit the high-colonic machine, manned by a fellow known among the staff as the Rear Admiral. Those patients who arrived to improve a specific aspect of their health—many came to lose weight—were assigned a personal practitioner who monitored their "diet, exercise and various vitality building procedures," Macfadden wrote in the November 1934 *Physical Culture*. A friendly but firm reminder appeared at the top of each individualized carte du jour in the dining room. "IMPORTANT: Your menu is checked for your own good. . . . Please do not embarrass the waitress by asking for something not marked as she has orders to follow the menu strictly."

Nude sunbathing was strongly encouraged, and men and women were provided with separate, walled zones on the hotel roof, inside which they partook of the healing rays "in the altogether," as one excited attendee put it. (Planes taking off from a nearby airfield would often circle above the hotel to get a bird's eye look at these heliotherapy sessions.) A network of trails latticed the hillsides, and visitors were encouraged to walk several miles each day, followed by dancing at night. A staff of four chiropractors dealt with the spinal aftershocks of these exertions. Saunas were strongly recommended, and could be followed by a dip in the hundred-foot-long spring-fed swimming pool. "Baths you take—all sorts of baths, according to your needs," was bon vivant Lowell Thomas's impression of the resort. "I gathered that anybody who comes away from Dansville is completely and absolutely clean, inside and out."

When Macfadden announced in 1935 that he would lead a 325-mile walk from New York City to Dansville, three major papers assigned reporters to follow its progress. On the morning of May 5, forty-eight hikers aged nineteen to eighty-three gathered on Broadway outside the Macfadden Publications Building. Former *Graphic* columnist Fiorello LaGuardia, now mayor of

New York City, fired a toy starter pistol. The press had dubbed the event a "Cracked Wheat Derby" because Macfadden was trying to prove that ordinary people could walk for two weeks on a simple, meatless diet rather than the steak-and-potatoes fare athletes were encouraged to consume. Behind the walkers came a truck carrying nearly a half ton of cereal, raisins, brown sugar, and salt. Next came a car pulling a trailer on which sat a piano and, according to *Newsweek*, a "plump torch singer, [who] banged the keys and led a chorus.

> **Cracked wheat our diet,**
> **Keeps tummies quiet.**

Other than dietary monotony and a few dropouts, the only serious problem occurred when the group arrived in Scranton, Pennsylvania, where two men who'd stripped down to their shorts were jailed for indecent exposure. They were freed and told to get out of town within twenty-four hours. A fortnight later, thirty-eight marchers strolled triumphantly into downtown Dansville, led by Macfadden. The winner of the weight-loss prize had dropped twenty-three pounds.

Macfadden's hotel, military academy, and health programs shared one theme in addition to *mens sana in corpore sano*: They all lost money. Describing his Castle Heights Academy in the early thirties, Macfadden wrote, "In all likelihood it will never be profitable from the point of view of dollars and cents. But while I have red on one side of the ledger, it is offset by the red-blooded men it sends into the world. Actuated by the right sort of ideals, they will be a credit to others and to their country." His next major purchase sought results of a greener shade. In June 1935, Macfadden took out a thirty-three-year lease on the oceanfront Deauville Beach Hotel in Miami, located on Collins Avenue where Sixty-seventh Street met the Atlantic. Macfadden spent $500,000 spiffing up the place, adding a third floor and nearly a hundred rooms and cabanas.

Advertisements that ran in national newspapers to announce the New Deauville's grand opening in January 1936 had a very different air than those for his other properties. Whereas Dansville's rechristening had required rounding up graduates of the old Physical Culture Training School, Deauville ads boasted that the facility was run by a former manager of New York's luxurious Hotel Astor. The drawing cards in Miami would all be comfort oriented: "Private Ocean Beach, 500 Feet Wide—Golf—Fishing—World's Largest Salt Water Circular Pool." Tucked into a fine-print box at the bottom of the ad was a small notice: "For those who wish to add to their store of health and vigor during their stay at the New Deauville, convenient arrangements may be made to follow Bernarr Macfadden's health-building diets and physical culture vitalizing methods." Should a brisk walk in the surf and a colonic cleansing fail to soothe one's nerves, less strenuous relaxants were available. A hotel staffer later recalled that the Deauville "had a bar on every floor," which "was neither a horizontal or a parallel bar."

The Roosevelts continued to write for *Liberty*, and to be written about in its pages. Franklin's speeches were reheated as essays, and Eleanor shared her opinions on subjects of interest to women. Their daughter Anna Roosevelt Dall maintained her office at the Sixty-fourth Street building after *Babies, Just Babies* folded. The Ourslers were frequent guests at the White House and at the Roosevelt home in Hyde Park, New York. Eleanor patiently responded to dozens of letters from Macfadden on the urgent need for a federally sponsored diet-and-exercise initiative and on other health topics. In one terse epistle, she politely explained that her husband "has had so many people who have remarkable cures visit him" that she doubted the president would take a consultation with Macfadden's friend Mr. Towns, who was evidently some sort of specialist in natural polio treatment.

But if the business and personal ties between New York and

Washington remained strong, the political one was starting to fray. Macfadden allowed Roosevelt a year to fulfill his promise to govern as a friend of business. When FDR revealed himself to be a populist in custom-tailored clothing, Macfadden's editorials in *Liberty* took on a nasty tone, knocking Roosevelt's National Recovery Administration and questioning the performance of the president's appointees. One source of Macfadden's antipathy was his self-made rich man's knee-jerk rejection of the New Deal, with its tolerance for welfare and taxes. Another was Macfadden's growing suspicion of communism. Some of FDR's closest advisors, including the future Supreme Court Justice Felix Frankfurter, were accused of sympathies pinker than a stack of *Graphic*s. The progressive who thirty years earlier had published *The Cry for Justice* was now printing editorials with titles like "Communistic Agitators in Our Schools—Hang the Traitors." Macfadden was far from unique among press lords in his feelings—William Randolph Hearst's empire of twenty-two red-baiting newspapers dwarfed Macfadden's influence—but his vehemence in fumigating the nation against red vermin was unparalleled. "Americanism is the only 'ism' I have ever dealt with," Macfadden was fond of saying (to the likely confusion of longtime vegetarians). So many of his editorials attacked communism that the Marxist weekly the *New Masses* paid him a compliment when it claimed, "For red-baiting, for opposition to anything socially progressive and for all-around viciousness, Macfadden can run rings around almost every other professional patriot in the business."

All the same, Macfadden pined for a government post. Despite his open campaigning for a federal health job, the highest rank he'd managed to attain in the Roosevelt administration was chairman of a fisheries committee, the purpose of which was to recharge the ailing U.S. fishing industry. After several meetings, the group's boldest recommendation was that Thanksgiving be moved from Thursday to Tuesday, so that Catholics accustomed to eating seafood on Friday wouldn't skip a week after the holiday. An invitation by Eleanor Roosevelt to present the Bernarr

Macfadden Foundation's Alabama youth fitness plan to her husband provided Macfadden a brief thrill—"I was pleased beyond measure when the President not only endorsed the idea, but made suggestions for extending it," he wrote to the First Lady afterward—but it ultimately came to nothing.

In the autumn of 1935, talk of Macfadden's political aspirations once again began to circulate. The source of this chatter was Bernarr Macfadden. At a meeting of the Republican Club of St. Louis, he let it be known that "he would not refuse the Republican Presidential nomination if it should be offered to him," reported the *New York Times*. The following month, he hosted a luncheon at his home in Hackensack for one hundred of New Jersey's top businessmen and Republican politicians, including the governor, at which a fiery speech about how FDR had jeopardized the Constitution was warmly received. He finally announced his intentions, sort of, in a long November interview with the *New York Herald Tribune*. The story's headline, BERNARR MACFADDEN EXPOSING SELF TO '36 PRESIDENTIAL BEE, should have tipped him off that not everyone was going to take his prospects for election as seriously as he did. Macfadden responded by denying his candidacy in *Liberty* just at the moment a glossy brochure titled *Bernarr Macfadden: Highlights of Fifty Years of Service for His Country* went on sale at newsstands nationwide. At Macfadden's prodding, the advertising agency that handled the *True Story* account begged radio stations to refer to their free-spending sponsor as someone "often mentioned for the presidency."

His undeclared campaign got off to a steady start. In Chicago, a large crowd applauded his attack on the New Deal. A Macfadden for President office, sponsored by Macfadden Publications employees and paper suppliers, opened on Michigan Avenue. An anti-Roosevelt speech before the Los Angeles County Republican Assembly in Beverly Hills earned approving front-page coverage in the conservative *Los Angeles Times*. "Huge amounts of money are being handed out with the nonchalance of a drunken sailor," Macfadden told the appreciative crowd. An opinion columnist inside the following day's paper commented, "Macfadden knows how to reach the masses and interest them, interest means influence and influence means votes. He can be of great aid in the restoration of sanity in Washington." A *Physical Culture* reader who skimmed the magazine's roundup of the top half-dozen Republican candidates for "the man-killing job" of commander in chief might not have been shocked to find among their number the one American whose fitness to lead the nation was beyond question.

Such flattering talk merely confirmed Macfadden's suspicions that he was just the man to rescue the Republican Party. This blind confidence made it all the more difficult for him to react when faced with people who did not entirely agree with that assessment. At a February speech before the Baltimore Advertising Club, which was carried on local radio, Macfadden addressed a favorite topic, "enemies that are within the secret chambers of our government and are the secret advisers of our Chief Executive." If he was expecting an outpouring of anticommunist support, he was surprised. "Keep politics out!" shouted a heckler. Macfadden started to talk about high taxes but was hooted into silence. The mayor of Baltimore asked that anyone who didn't want to hear the speech get up and leave. No one moved. Macfadden turned to the topic of the Yellow Peril and declared that 250,000 armed Japanese were living in California, preparing to attack the United States. The room erupted, and the speech ended abruptly.

Macfadden continued to tour through the spring of 1936, drawing large crowds and often using his forum to mix health advice with campaign platitudes. MACFADDEN RIPS INTO NEW DEAL AND BIG FEEDERS was the *Chicago Tribune* headline over a story about one well-received speech to 1,800 people. All along, he maintained his nebulous position that while he wasn't quite a candidate for president, he would, like any good American, feel patriotically bound to accept the nomination should it be thrust upon him. In June, Macfadden tipped his hand slightly to a *Tribune* reporter in Cleveland on the eve of the Republican convention. He admitted that since no candidate had arrived in town as the clear-cut winner, he was planning to enter a deadlocked convention as a dark-horse nominee. "Off-hand, I'd estimate my support at 200 to 300 delegates," he told the reporter.

Front-runner Alf Landon had locked up 179 of the convention's 1,001 delegates. What made Macfadden so confident that he could stand on the sidelines, ready to pounce, with an even greater number safely in his pocket? Almost half a century later, Macfadden Publications' onetime circulation director S. O. Shapiro admitted to Macfadden biographer Robert Ernst that his boss had given $100,000 to a consortium of politicians from the Midwest. For that price, they promised, "he'd get control of a number of delegates who would be uncommitted for the first few ballots and would promise to swing to B.M. when the convention was deadlocked," remembered Shapiro. Macfadden Publications executives who pointed out to their boss that such a scenario sounded improbable, at best, were cut out of further discussions. "When the convention took place," Shapiro said, "B.M. sequestered himself in his hotel room in Cleveland with a radio to hear how the whole thing came off. He ordered me to keep him company. He sat there crestfallen when his name was never mentioned."

Chapter 15

Macfadden for Senate

Are We to Become a Nation of Sissies?

—TITLE OF BERNARR MACFADDEN'S EDITORIAL IN *LIBERTY*,
SEPTEMBER 18, 1937

Macfadden Publications' stock price fared better than its founder's political fortunes, marching back from the low teens to the mid seventies. Macfadden's spending spree had netted some smart purchases, including the Hollywood movie magazines *Photoplay* and *Shadowplay*. Oursler's stint in Hollywood had ushered in a new wave of celebrity coverage on Sixty-fourth Street. Rare was the month in which *Physical Culture* had nothing nice to say about Norwegian skater-turned-starlet Sonja Henie. Kindergarten fitness buffs surely took inspiration from "Shirley Temple's Diet and Training." *True Story*'s circulation, which had dipped during the bleakest years of the Depression, once again broke 2 million copies per month and continued to rise, driven by articles such as "The True Love Story of Greta Garbo," in which the reclusive star "saw youth and love for the first time ... and she knew that she was lost." Other investments were less successful, such as a short-lived magazine called *Your Faith*, which was advertised not with the face of Jesus or Muhammad, but rather that of a certain bushy-haired health prophet. *Liberty* continued to churn out its fascinating mix of high and low culture; one span of a few weeks included contributions Oursler had coaxed from H. L. Mencken, Leon Trotsky, and formerly bitter *Graphic* graduate Walter Winchell, as well as Macfadden editorials such as "Billions in Additional Taxes Due to Official Incompetence." The weekly suffered a drop in topicality—and sales—during 1935, when Oursler took a long sabbatical to Egypt and Palestine, but it came back strong in 1936, thanks to two multipart stories the editor engineered. The first, inspired by a visit with President Roosevelt, was a serial entitled "The President's Mystery Story." FDR, an aficionado of whodunits, told Oursler of a plot that had been rattling around his brain for years: How could a wealthy man walk away from his life with five million dollars and never be traced? Oursler assigned the plot to several top mystery writers, who fleshed out the details, unaware of who'd drawn up the story outline until the tale was published in an issue with the president's smiling face on the cover.

The second serial caused an even bigger sensation. The Lindbergh baby kidnapping and murder case had been solved, and Bruno Richard Hauptmann had been convicted of the killing. The events leading to his capture had been unusual, to say the least. When the police investigation had stalled, Dr. John F. Condon, an eccentric college lecturer, had the idea to place a classified ad in the *Bronx Home News* offering to serve as an intermediary between the kidnapper and the Lindberghs. Such proposals were common. (During the dying days of the *Graphic*, Macfadden had repeatedly offered his services to authorities.) Unlike all others, Condon had received a positive response in a letter adorned with a cryptic symbol. Charles Lindbergh confirmed that the sign had appeared on a ransom note left in his son's room. Condon (who used the name "Jafsie" in these dealings) arranged a meeting with the stranger in Woodlawn Cemetery, and followed the man's instructions to return with $50,000. Weeks later, the baby's body was uncovered in a shallow grave near the Lindbergh home. Hauptmann, a German-born carpenter who lived in the Bronx, was captured when he tried to pass one of the bills given as ransom.

Aside from the Lindberghs, who escaped abroad to avoid the crush of press attention, and Hauptmann, who maintained his innocence from prison in New Jersey, Condon was the most intriguing player in the kidnapping drama. Oursler invited Condon to Sandalwood to dictate his recollections for a book, *Jafsie Tells All*, which was serialized in *Liberty*. The January 22, 1936, *Liberty*, which contained part two of Condon's memoirs and featured the Prince of Wales on the cover—at the moment of his coronation as King Edward VIII—sold more than 2.6 million copies, the greatest number in the magazine's history.

The publication of Condon's tale was not the end of *Liberty*'s, or Macfadden's, involvement in the Lindbergh case. Hauptmann's execution was stayed, to the consternation of the entire country, in part because New Jersey governor Harold Hoffman felt that the Jafsie series had raised doubts about Hauptmann's guilt. In the wee hours of March 13, Hoffman paid a visit to Oursler's suite at the Waldorf-Astoria Hotel. After discussing the case for "some time," Oursler recalled, the governor told him: "You seem to be the kind of person I can tell the truth to. Did you know, Mr. Oursler, that you have been accused of murdering the Lindbergh baby?"

Oursler laughed and said that he'd "been accused of everything else in my life but this."

"Don't laugh," replied the governor. "Wait until you hear the source of the accusation. I have a letter pointing out that you were the mastermind who inspired the crime in order to get the Jafsie articles for *Liberty*. The person who wrote this letter and who wrote very bitterly about you and Mr. Macfadden was Mrs. Mary Macfadden."

The note of which Hoffman spoke accused Oursler of being a "magician and hypnotist" who preyed upon Bernarr Macfadden's love of publicity and had conspired "to take and hold for ransom the Lindbergh child but without intent to kill or harm it." For Oursler, the double upside of committing this crime would, the letter's writer explained, be "the permanent gratitude of my husband," who would earn national acclaim for under-writing the child's return. Oursler, in turn, would "also be the recipient of a large reward paid by Mr. Macfadden."

Mary denied having written the note and claimed that it was a forgery sent to embarrass her. She did admit, however, to having sent two previous letters to Hoffman's office that had painted Oursler in similar negative terms—calling him "one of the clever-est charlatans I have had the displeasure to meet"—and had hinted that he might be involved in the Lindbergh case. Her lawyer com-plained that the letters had been sent to Hoffman in confidence.

In early 1937, Oursler sued Mary Macfadden for libel, asking $150,000 in damages. Bernarr Macfadden at first seemed to sup-port the suit, writing to Oursler that he "would consider it a big favor if she were properly penalized regardless of the publicity." As the trial approached, though, Oursler dropped the charges. *Time* later insinuated that Oursler had arranged the whole ruse himself by paying someone $100 to send the accusatory note. The source for this charge was likely Mary's lawyer, who was quoted in the same article as saying that his client planned to press criminal charges. She never did.

The approaching war in Europe surely stirred conflicting emotions in Macfadden. His *Liberty* editorials waved the flag with all of their author's considerable might. If anything could prevent him from getting his mandatory eight hours of sleep at night, it was worrying about the preparedness of America's armed forces for battle. When President Roosevelt announced an ambi-tious plan to build fifty thousand warplanes, Macfadden de-manded that the country build one hundred thousand. The Army Air Corps—soon to become the U.S. Air Force—was having trouble composing a fight song to rival the Navy's "Anchors Aweigh," so Macfadden sponsored a $1,000 prize that inspired the anthem that begins "Off we go / Into the wild blue yonder." *Liberty*'s far-flung correspondents were dispatching stories from an increasingly anxious Europe. The Princess Catherine Radziwill voyaged to Russia and returned with what she claimed was an interview with her cousin, Joseph Stalin, in which he said he was considering an alliance with Hitler. *Liberty* printed her story, which the Kremlin denied, and was soon vindicated by events. Cornelius Vanderbilt filed reports from every corner of the continent (while also secretly operating as Roosevelt's personal spy); he once returned to his Danzig hotel room to find SS Commander Heinrich Himmler inside, listening to the news on Vanderbilt's radio.

Yet deep in his heart, the love Macfadden felt for his country wrestled with his love of physical culture. And the two nations on earth that he revered above all others for their efforts in producing vigorous men through proper breeding, diet, and exercise—surpassing even Mussolini's Fascists, whose sculpted, outstretched arms were quietly airbrushed from Macfadden magazines as the thirties continued—were Japan and Germany, the very countries that posed the greatest potential threat to American security.

Early in the century, Macfadden had reckoned that the Japanese, fresh from crushing the Russians in their war of 1904–05, could thrash the United States in combat. "As a race they possess more strength and endurance per pound weight than any other race in the world," he wrote in *Physical Culture* in 1907. Not only did they exercise regularly, but their low-calorie, largely meatless diet was an example for the rest of humanity. Macfadden's alarm at a quarter-million Japanese lurking in California*

On the night of the Pearl Harbor attack, FBI Director J. Edgar Hoover— never a stickler for civil rights—estimated the number of Japanese in the United States who might be rounded up for questioning at 770.

was probably heightened by this subconscious fear of their superiority.

By the mid-1930s, the resurgent Germany was growing belligerent. Though Macfadden was hardly an anti-Semite and promoted Jews to top executive positions, the German martial ideal held an unshakable grip on him. (George Sylvester Viereck, a Nazi propagandist whose pro-German sympathies had put his photo on the front pages of New York newspapers during World War I, served as one of Macfadden's closest advisors until his subpoena for un-American activities in 1938.) Macfadden never wavered from the feelings he poured into this 1917 editorial about how one nation was nearly able to fight all of Europe to a draw in World War I:

With but little more than half the population of the United States, the Teuton race has sustained the most gigantic military struggle of history. . . . To what does Germany owe her extraordinary military power? It comes and can only come from the physical resources of her people. The secret of the Teutonic achievements is to be found in two words: PHYSICAL CULTURE.

Two decades later, in an editorial titled "Democracy Is Doomed, Says Dr. Goebbels," Macfadden bemoaned his own country's failure to promote "efforts to build a race necessary to maintain the vigor and vitality so badly needed to defend our democracy." One can only wonder what passed through Macfadden's mind as he watched newsreels of Germany's vegetarian dictator flaunting his racially pure athletes at the 1936 Olympiad. Macfadden maintained an isolationist position in *Liberty* almost until the attack at Pearl Harbor, while simultaneously publishing story after story in *Physical Culture* about the ominously sorry shape of America's fighting men. A note sent to Eleanor Roosevelt promoting his "walk to work" campaign in 1940, when swastikas were flying over Paris, had an "I told you so" ring to it. "The thousands of walking clubs throughout Germany undoubtedly

[have] much to do with the vitality and vigor which developed the fighting spirit of the Germans," he explained to the First Lady.

One side effect of being the world's most famous health guru, as well as the employer of a very large publicity department, was that Macfadden's name often cropped up in newspaper gossip columns. In November 1939, the nationally syndicated columnist Leonard Lyons ran an item that said, "For the first time in the many years since he's been wintering in Florida Bernarr Macfadden will establish a legal residence there December 15." Macfadden worshipped the sun, of course, and the Miami Deauville required an increasing amount of his attention. But in 1940 Florida possessed an added enticement—a U.S. Senate race in which the incumbent was potentially vulnerable.

Florida in the last year before America entered World War II was only a sleepy hint of the behemoth it has become. It wasn't an especially prosperous state, having suffered deeply during the Depression, following a Jazz Age land boom and bust and devastating hurricanes in 1926 and 1928. Nor was Florida particularly influential politically. Its population ranked twenty-seventh out of the forty-eight states, well behind both Iowa and Mississippi. Cigar rolling was a major industry; Pan Am flew regularly to four cities in Cuba; any oversized mice seen in Orlando were there to sample the produce of the citrus industry, not welcome visitors to Space Mountain. To Bernarr Macfadden, it seemed the sort of place where a political candidate might win votes based almost solely on name recognition and reputation.

When writing his reminiscences of the senate campaign, Macfadden provided a questionable rationale for his entry into Florida politics: He was hosting a meeting of Macfadden Publications executives in Miami in March of 1940, and, lo and behold, one of them suggested that he run for the senate. (The idea that an underling might have been kissing the boss's gluteals, still capable of vaulting him over a Deauville deck chair from a standing start at age seventy-one, seems not to have occurred to Macfadden.) He would run as a conservative Democrat. Florida was a one-party state—the Republicans didn't even bother to enter a candidate in that same year's gubernatorial election—so the winner of the Democratic primary was almost guaranteed a six-year trip to Washington, D.C. The contest was to be held on May 7, and aspirants needed to file by March 23. Macfadden placed ads in the major Miami newspapers inviting voters to attend "an open forum discussion," at the Deauville on March 22, with Oursler as master of ceremonies. The topic was whether Macfadden should run for the senate. "From the time of my entrance until the meeting was over the idea met with such a favorable response that procedure became merely a matter of form," Macfadden later wrote. The next morning at dawn, he took off from Miami Municipal Airport in his Bellanca monoplane, flying the seven hundred miles to the state capital at Tallahassee to pay his $500 filing fee. "With my northwesterly course set to take me over the vast stretches of Florida's great Everglades and into its rich citrus belt and on over the great prairie lands of the cattle country," he wrote, "I felt a newer and greater sense of exaltation—a feeling that more clearly manifested itself with a deep conviction that I was about to enter a phase of my life which bore possibility of reaching its greatest accomplishment."

The state's political reporters, greeted unexpectedly with an eccentric celebrity candidate who planned to barnstorm around the state in his personal airplane, were understandably cheered. Prior to Macfadden's appearance from the sky, they'd been faced with a contest chiefly between three native sons: bland incumbent Senator Charles Andrews, tobacco-chewing Governor Fred Cone, and backroom deal-cutter Jerry Carter, head of the state railroad commission. All were New Deal supporters. Florida's election laws mandated that if the leading vote getter in the primary didn't capture at least a 50 percent majority, a runoff was to be held between the top two finishers. Macfadden was betting that if he could run a strong second to Andrews, he might be able to unite the state's Republican minority and anti-Roosevelt Democrats in a winning coalition.

Editorial writers were somewhat less enthused. "I soon learned that I was a black Republican and a Yankee interloper," Macfadden recalled, "a Croesus rolling in wealth and waxing in the luxury of a 'palatial' Miami Beach Hotel." On the first day of business at Macfadden for Senate campaign headquarters, so many representatives of special interest groups arrived open-handed offering their support," Macfadden wrote, that "if we had paid the sums of money they asked for I would have received a total vote twice the population of the State of Florida."

More troublesome than any carpetbagger image problems were the logistical issues. Six weeks before the primary, the state's major press endorsements had been locked up. Macfadden had no advance team of local political professionals to plot his campaign strategy, only the aid of publishing-industry staffers from New York City. Their pushy tactics and Henny Youngman accents appalled Florida's political elite.

Macfadden also hadn't given much thought to his political platform, beyond slenderizing federal spending, taxes, and the spread of "isms," while strengthening just about everything else: the military, states' rights, the spinal musculature of Floridians. He was certainly the only candidate who pledged to promote the entire state of Florida as a health paradise. "All kinds of diseases fade away like snow in the summer when people come here," he wrote in one campaign brochure. "Nearly everybody up north suffers with arthritis and sinus trouble. . . . Even a few days will often drive these complaints away in the Florida sunshine." He promised that his magazines—the top sellers on America's newsstands—would sing the state's praises in every conceivable manner.

Macfadden crisscrossed the peninsula like a bootlace, piloting himself to well-attended speeches in Jacksonville, St. Petersburg,

Daytona Beach, West Palm Beach, and Ocala. On April 13, half-way through his whirlwind campaign, Macfadden landed in Tampa. During a talk with reporters, a writer from the *Tampa Tribune* grilled him about his party affiliation. The paper followed with an editorial demanding Macfadden's answers to a list of questions:

> Are you registered as a Democrat in Florida?
> If so, how long have you been registered?
> If you claim to be a Democrat, state and national, when did you first make that claim?
> Have you ever voted a Democratic ticket in Florida?
> Were you a Democrat or claiming to be a Democrat in the national election of 1936?

Macfadden hastily replied that he considered himself a "Jeffersonian–Andrew Jackson Democrat" who had heartily supported Roosevelt until "he surrounded himself with pink-colored theorists and socialistic politicians." He generously gave himself credit for two years of Florida citizenship.

Macfadden had walked into a trap. The *Tribune* promptly noted that the Florida state constitution prohibited certifying the election of any candidate for the Senate who had not been a state citizen for at least five years. "In the face of that," wrote the paper's editorialist, "Mr. Macfadden would be eliminated, even if he was elected."

Macfadden fumed, believing he could win a court battle against such a law, but aware that launching a counterattack would broadcast his shaky claim to citizenship. "No voter wants to cast his vote for a certain loser," Macfadden knew. He ignored the issue and increased his spending on advertising. His team, now bolstered by a few local politicians, concentrated its efforts on publicizing a Macfadden speech scheduled in Miami on April 20. Newspapers were stuffed with advertisements. Mailboxes groaned with muscular propaganda. Sound trucks patrolled the streets. The campaign leased two airplanes and rigged them with enormous public address systems. Within days, the residents of Miami Beach and other cities around the state were hounded by a disembodied voice commanding them from above to "Boost Florida with Macfadden!" Resistance was futile. "Our sound airplanes, I suppose, provided about the most spectacular type of campaigning ever witnessed in Florida," Macfadden later recalled proudly.

On the evening of his Miami rally, five hundred *Liberty* carrier boys and a team of drum majorettes led a torchlight parade through the city to Bayfront Park. Macfadden's advance team had promised reporters "an announcement of intense interest to every citizen in Florida." At least seven thousand people turned out to hear Macfadden's speech, which was carried live via radio around the state. His big surprise was that he would forego his first year's salary and use those funds to establish a publicity department for Florida tourism. To modern ears, the shocking revelation is that Florida's Precambrian vacation industry had no advertising budget. As far as Macfadden was concerned, his sacrifice was a masterstroke—he gauged its effect as "quite electrifying"—that put the race's momentum squarely behind him.

Two days later, a news item of indisputably intense interest came across the Associated Press wires.

NEW YORK, April 22 (AP)—A suit started in Federal Court today against Bernarr MacFadden and MacFadden Publications, Inc., for an accounting of $1,400,000 alleges that MacFadden spent $250,000 of the corporation's funds in 1936 in an attempt to obtain the Republican Presidential nomination.

Macfadden saw the legal action as an attempt by his rivals to mitigate the impact of his Miami speech. "Nothing else but political animosity could cause such a suit to be brought after four years of silence," he told the *New York Times*. "The books of

Macfadden Publications are audited by one of the largest and most reputable auditing organizations in the world." Then he dropped the subject.

In the final week of campaigning, the Macfadden for Senate team mailed circulars to every voter in the state, and *Liberty* published an "All-Florida" issue. The tireless candidate hopscotched to fifty speaking engagements up and down the state. Three thousand college students in Gainesville, one stop on a six-hundred-mile circuit that day, cheered his exhortation to seek "health in their government, as well as health in their bodies." His final push ended with speeches in the naval strongholds of Key West and Pensacola, at opposite ends of the state. On May 7, election day, he flew back to Miami, sneaking in a long power walk during a refueling stop. He cast his ballot and waited impatiently in his room, then departed for his campaign headquarters shortly before sundown.

"When I arrived about seven o'clock in the evening the crowd had overflowed into the street," Macfadden recalled afterward. "Early returns showed Macfadden in the lead! A miracle was in the making." Results from the Miami metropolitan area, where voting machines were in wide use, were among the first ballots tabulated. "Headquarters was a bedlam," Macfadden said. Once-frosty politicians, sensing a change in the wind, elbowed in to offer congratulations and support. Reporters begged for a victory statement. A local radio station asked Macfadden to comment live "on the most remarkable political upset in Florida's history."

As expected, Senator Andrews soon took a large lead over the pack, but he was projected to fall far short of the majority necessary to prevent a runoff. When Macfadden went to sleep on the floor of his Deauville apartment after midnight—a firm surface being better for the back, he believed, than a soft bed— almost half of the state's precincts had reported. Macfadden owned a 3,500-vote lead over Carter and 13,000-vote cushion on Cone.

The next morning was one of the greatest of Macfadden's life. He picked up a copy of the *Miami Herald* and saw the headline:

ANDREWS ON TOP, MACFADDEN 2ND
Physical Culturist's Race Surprise; Cone Is Given Bad Defeat

The eyesight of the author of *Throw Away Your Glasses* had deteriorated with age, so he must have held the paper at arm's length in order to read the *Herald* political editor's postgame analysis. "The story this morning is that of Bernarr Macfadden, the amazing man nobody in Florida politics gave a chance in the senatorial race. Unless an unexpected shift develops, the 72-year-old [actually, 71] messiah of health is to be the runnerup Senator Andrews must beat on May 28 to remain in the senate."

The events of May 8 took a turn shortly after Macfadden read those words. "The switchboard at the Deauville was constantly cluttered with 'important,' 'urgent' and 'emergency' calls for me," he recalled. According to Macfadden, one of the other candidates—whom he refused to name, but who was obviously Jerry Carter—whose "personal machine" controlled precincts in the northwest part of Florida, was holding Macfadden votes hostage for ransom. This candidate, Macfadden later argued (not very convincingly), had entered the senate race not to win, but in hopes of selling his support to *another* candidate in the case of a close outcome. Macfadden instructed a staffer to deliver his response to the mysterious opponent's offer: "Tell him to go to hell."

A smaller May 9 *Herald* headline was as deflating as the prior day's had been exhilarating:

**MACFADDEN DROPS BACK AS WEST FLORIDA VOTES
MAKE CARTER RUNNER-UP**

A few days later, Carter's second-place finish became official. Macfadden placed third, nine thousand votes behind Carter.

Andrews demolished his opponent in the runoff and returned to Washington. Macfadden, conveniently forgetting certain transactions he'd made to buy convention delegates a few years before in a Cleveland hotel room, congratulated himself on maintaining his honor. "Like Shakespeare's Hamlet, I had a choice to make," he wrote in his campaign memoir. "Mine was to win or not to win. I chose not to win because in such a triumph, although it involved only the release of votes I believed to be rightfully mine, and might well have resulted in my ultimate election to the senate, I could have considered it only an empty victory."

For a man as certain of his destiny as Bernarr Macfadden, nearly earning a man-to-man shot at the U.S. Senate in just six weeks only affirmed his belief that he belonged in public office, no matter the cost. But before he could plan his next political move, there was a small mess to clean up—that shareholders' lawsuit back in New York.

While Bernarr Macfadden struggled to get his political career airborne, the influence of physcultopathy flourished as never before. The surprisingly warm response to his Florida campaign had proved that a man could be taken seriously after forty years of flashing his pectorals and hectoring the nation to monitor its bowel movements. The self-professed "physical-culture crank" had persevered long enough to see many of the ideas for which he'd fought seep into American life. And he wasn't going to shut up about it, either. "Anybody who will take the trouble to look back through the yellowing pages of those early issues of PHYSICAL CULTURE as it was at the turn of the century will find ample evidence on one point," the magazine boasted in its fortieth anniversary issue of December 1938. "That is that Bernarr Macfadden was, and still is, one of the prophets of his day. For practically everything that he then advocated in the way of common-sense change and reform in ways of living, and in the prevention and treatment of disease, has either come to pass or is evidently on its way."

A spread of photographs in that same issue of *Physical Culture* contrasted women's swimwear in 1898, in which ladies waded daintily into a lake wearing what looked like long-sleeved bridesmaid's dresses, with modern form-hugging one-piece suits that accentuated every feminine curve. The Macfadden ideal of male perfection was now so prevalent in Hollywood—this was the era of Johnny Weissmuller's Tarzan—that leading men routinely sat down with Macfadden for *Physical Culture* interviews to boast about their fitness secrets. (Clark Gable usually ate a light breakfast; Jimmy Cagney, then appearing onscreen as the merciless killer Rocky Sullivan in *Angels with Dirty Faces*, told Macfadden, "I have some exercises of my own that I think are pretty effective for the buttocks and the stomach.") The first Mr. America contest was held in 1939, helping to usher in the age of weight training for purely aesthetic reasons. The young muscle promoter Joe Weider, architect of the bodybuilding-as-entertainment industry, was awed when Macfadden summoned him to lunch at the New York Athletic Club and predicted his future success. "I felt like God had anointed me," Weider recalled. Just as his idol had discovered Charles Atlas, Weider found a strapping, ambitious immigrant of his own—Arnold Schwarzenegger.

As Hitler's armies moved through Europe at will and British soldiers returned from battle telling terrifying tales of titanic Teutons—while forty percent of American men called up for military service starting in September of 1940 were turned away for medical reasons—even Macfadden's call to conscript all American men aged sixteen to forty into an exercise program sounded almost sensible. (Macfadden's own eugenic dreams had faded, but confirmation of his breeding and training theories could be found in his youngest son Brewster, a Yale swimming champion. In early 1943, before shipping off to the Pacific with the Navy, Brewster swam the fastest legs on two relays that smashed world records.)

Macfadden's position as America's leader of the rapidly growing movement that he called Eating for Health was cemented in May 1938, when *Physical Culture* was named the official publication of the National Health Foods Association. While brown rice and mung beans were still cultish—the late 1930s were the golden age of canned foods, automat lunches, and Betty Crocker cookbooks—the groundwork was being laid for the health food resurgence of the 1960s. In June 1939, the *New York Times* ran a front-page story trumpeting the findings of a Rockefeller Foundation scientist who fed a thousand lab rats a restricted diet based on the eating habits of peasants in northern India, "among them some of the finest specimens of mankind," according to the study's organizer. Not a single rat fell ill in two and a half years. A thousand other rats were fed a diet similar to that of the average American; these animals suffered from thirty-nine different diseases almost identical to those that plagued overfed humans. The message was clear: to stay healthy, eat less, and eat simply. "At last medical science has acknowledged the truth of the dietetic principles that I have been teaching as a propagandist for more than 50 years," Macfadden boasted in a *Liberty* editorial.

The American Medical Association, while it never slackened its condescending attitude toward Macfadden, had also begun to inch in his direction regarding nutrition. In 1938, the association published "The Normal Diet," a sensible set of eating guidelines. Late the next year, *JAMA* published an editorial recommending that processed foods should be fortified with sufficient vitamins and minerals to recapture their "high natural levels." With rumors swirling of Hitler using vitamin B_1 as a secret weapon, America's white-flour millers agreed to add thiamin and other nutrients to their product, creating the varieties of "enriched" flour and white bread still found on supermarket shelves. To prove the wholesome healthiness of this new bread, the bakers even advertised in former enemy territory, *Physical Culture*, and other Macfadden magazines. Macfadden told an interviewer that the reenrichment of the nation's white flour was "the dream of my life come true."

Around this time, Alva Johnston, one of America's premier journalists, wrote a long, flattering profile of Macfadden that appeared in *Liberty*'s dominant rival, the *Saturday Evening Post*. In taking the measure of Macfadden's impact on America, Johnston tried to calculate Macfadden's "lifelong batting average" as a reformer:

Against prudishness	1.000
Against medicine	.000
Against corsets	.890
Against muscular inactivity	.333
Against alcohol	.250
Against cigarettes	.000
Against white bread	1.000

"This gives him a grand batting average of .496," Johnston wrote. "A mark seldom equaled by any crusader."

As a businessman, though, 1940 was a year of nothing but strikeouts for Macfadden. *Liberty* was getting squeezed by new competition. Henry Luce, the publisher of *Time* and *Fortune*, had turned the weekly magazine market on its head by introducing his picture magazine, *Life*, in 1936. Almost instantly *Life* was selling a million copies per week. While *Liberty*'s salesboys were still knocking on doors and pushing their magazines for a nickel per, readers were begging their local newsdealers to save them ten-cent copies of *Life* because many issues sold out. The biweekly photo journal *Look* appeared in 1938 and, like *Life*, quickly found an enormous readership. Advertisers followed. The outlook for 1941 was dismal. The company was printing 2 million more copies of its magazines than it had in 1935, but the net sale of those titles had declined by 1.5 million copies. *True Story* sales, especially, were flagging again, and *Physical Culture*'s influence far outstripped its income; Macfadden's beloved magazine was hemorrhaging money. Shares of Macfadden Publications stock had dropped ninety percent in four years, to just $7 per share. Dividends had fallen so seriously in arrears that the company would soon owe an incredible $28.50 on each share of its preferred stock.

By the time Macfadden returned from his Florida campaign, the shareholders' lawsuit against the company's majority stockholder had been bundled with two others into a single complaint, *Leon S. Brach* v. *Macfadden Publications Incorporated and Bernarr Macfadden*. Macfadden was charged with bullying executives and using the company treasury as his personal piggy bank. He gave free advertising space to properties owned by the Bernarr Macfadden Foundation, the suit contended, and on occasion steered company cash to Castle Heights and the Miami Deauville. His $50,000-a-year salary was deemed excessive for what was at best part-time work. All told, Macfadden stood accused of diverting $900,000 in company funds for his own use, and with trying to hide $2.9 million in losses suffered by the *Graphic*. These last were criminal charges. If found guilty, Macfadden would go to prison.

What Macfadden didn't know was that the suit he first imagined to be the work of political foes was in fact a mutiny. Leon S. Brach was a straw man. The lawsuit was engineered by some of Macfadden's most trusted executives. Vice President Orr J. Elder, who'd started with Macfadden nearly forty years earlier as assistant advertising manager of *Physical Culture*, had come to view the company's founder as a liability. The Lady Macbeth behind the regicide plot seems to have been the company's longtime corporate attorney, Joseph Schultz. (Schultz and Elder, along with Oursler and Susie Wood, had been founding members of the Bernarr Macfadden Foundation board of directors.) Macfadden had fired Schultz a few years before for refusing to take a pay cut. The lawyer had kept records of Macfadden's financial transactions. "I can prove that he's taken several hundred thousand dollars from the company," Schultz told the head of a competing publisher. No one can say for certain, but Orr Elder's son believed that his father decided to join Schultz and other executives as he watched the company's fortunes decline following the 1936 election. After begging Macfadden to cut spending and refocus on his primary business, to no avail, Elder and his fellow executives decided that orchestrating a coup was their only hope for a turnaround.

Macfadden's legal defense was that, sure, he'd floated his name for the presidency in 1936, but that he'd done so only for the benefit of Macfadden Publications. The directors, he wrote to his lawyer, Arthur Garfield Hays, in late May 1940, knew "that it would probably be worth half a million dollars or more to the stability of the company to have the Macfadden name prominently mentioned for this high office . . . it was the general conclusion that the associated publicity was well worth the expenditure." Company officers had signed off on every financial decision. To the end, he insisted that he'd done nothing wrong. "Macfadden never seemed to realize that he had a public company and that there were some things he couldn't do," said one former executive.

The fight was waged behind closed doors for the better part of a year. Almost four decades later, Fulton Oursler Jr. met a former Macfadden Publications employee who swore that Mary Macfadden had hired private detectives to follow her husband. According to this version of the tale, the investigator photographed Macfadden with various women in various cities, and Mary had copies of these pictures—some of them explicit—delivered to the boardroom before company directors arrived for their first meeting on Macfadden's fate. When Oursler Jr. asked S. O. Shapiro, who had left the company by the time of the lawsuit but who remained in close contact with its executives, whether he thought the story was true, Shapiro refused to confirm or deny it.

On February 27, 1941, a settlement was announced. Macfadden Publications would buy 22,162 shares of voting stock back from Macfadden. He, in turn, would pay the company $300,000. Forty-two years after first declaring in *Physical Culture* that weakness was a crime, Bernarr Macfadden severed ties with his namesake company to avoid jail time. Beginning with the June 1941, issue, the bold signature of *Physical Culture*'s founder disappeared from its opening page. Less than a year later, the magazine was out of business.

Chapter 16

Paradise Lost

Nothing that I have ever read of Macfadden, including that which I have written myself, has ever captured him.

—FULTON OURSLER, *BEHOLD THIS DREAMER*

Like any autocrat forced to abdicate, Bernarr Macfadden expressed a sudden desire to live in exile. "Having reached the age of 73 and finding that my Foundation Enterprises are demanding more and more of my time and attention," he announced to reporters, "I have decided to relinquish control of Macfadden Publications, Inc." Privately, he was devastated and feeling unmoored for the first time since he'd left St. Louis. He walked the streets of Manhattan, mile after mile after mile, in silence. "The position of your father has changed quite considerably since his retirement from the publishing business," Macfadden wrote to his teenage son Brewster in the spring of 1941, seven weeks after his ouster. "With the radical reduction of my income . . . a radical reduction will have to be made in my expenses."

A black cloud seemed to follow Macfadden through the spring. In March, the Federal Trade Commission ordered the Macfadden Book Company to stop selling his popular *Hair Culture*, a manual devoted to his hair-pulling system of follicular rejuvenation, which the FTC said fell short of its claims to "prevent, correct the cause of, or cure, baldness, dandruff, brittleness, split hair or graying hair." Weeks later, an internal Macfadden Publications audit uncovered the scandalous secret that *True Story*'s and *Liberty*'s circulation numbers had been inflated by tens of thousands of copies in 1940. Newsstand agents had been promised a bonus if they maintained sales quotas, and apparently the sweetener had been tasty enough that many agents, rather than returning unsold copies, tossed them in the trash to collect their premium. Advertisers, who then as now paid according to the carefully measured size of a medium's audience, were not amused. The new executive team at Macfadden Publications, some of whom whispered that Macfadden might also have taken kickbacks from suppliers, made clear to Madison Avenue that the party responsible for these despicable financial shenanigans was no longer with the company.

Although Macfadden remained living proof of his own antiaging theories, his outward vitality—newspapers nationwide published photos of him standing on his head atop his desk on his seventy-fifth birthday—masked an unfortunate truth: His legendary sense of timing and public taste had atrophied. After Macfadden Publications euthanized *Physical Culture*, Macfadden's former subordinates were only too happy to sell the money-losing magazine back to its founder for a nominal sum. (This in spite of a five-year noncompete agreement he'd signed as part of his settlement.) *Physical Culture* was reintroduced as a digest-sized publication in 1943. By that time, Americans had turned their attentions to more pressing threats than the medical monopoly.

Macfadden boasted of plans to start a new publishing company with sons Berwyn and Brewster when his five-year hiatus expired. In the interim, his full-time job was overlord of the Bernarr Macfadden Foundation's properties. Aside from the Dansville hotel, which maintained a steady business, those facilities did nothing to threaten the foundation's not-for-profit status. The Loomis Sanitarium, a hospital that Macfadden had purchased in 1939 and spent $90,000 on renovating as a physcultopathic treatment center for tuberculosis, was closed in 1942. The

U.S. Army commandeered the Deauville the same year. After the war, the hotel was returned, slightly bruised and saddled with a new problem. In his rush to take possession of the Deauville, Macfadden had signed a somewhat ambiguously worded thirty-three-year lease, which the hotel's owner (or rather, her new husband) now interpreted as entitling her to half of the resort's *gross* receipts—retroactive to 1936. What with legal fees, upgrades, and repairs from hurricane damage in 1947, Macfadden sank a million dollars into the Deauville before getting evicted in 1949, just in time for the Florida tourism boom of the 1950s. Meanwhile, the enormous piece of investment property that Macfadden had ceded to the state of New Jersey because of a half million dollars in taxes was being developed into millions of dollars' worth of homes for postwar families. Today, residents of Bergenfield can take a left off Graphic Boulevard onto Elder Avenue on their way to visit neighbors on Brewster Place, but names are all that remain of Macfadden's investment.

If Macfadden instantly grasped the changes coming from his loss of income, he was much slower to intuit a second shift—that while a wealthy media mogul who walks to work barefoot with a forty-pound bag of sand on his back is a charming eccentric, the president of a small foundation who acts in such a manner is a social liability. With the loss of his publishing company, many of Macfadden's influential friends were now tuning him out. In the late 1930s, *True Detective*'s most enthusiastic booster had been FBI director J. Edgar Hoover. Fulton Oursler spoke at the bureau's graduation ceremony in Virginia; Hoover returned the favor by appearing at *Liberty*'s Award for Valor in Citizenship luncheon in Manhattan. "It has always been a pleasure to cooperate with your organization," Hoover wrote to Macfadden in early 1940. "We are both charged with the upholding of law and order." To Oursler, Hoover was even more effusive: "Very frankly, I consider you as a member of the organization."

Five years later, in 1944, when Macfadden was trapped in Miami desperately fighting off his landlord's lawsuit at the Deauville, he sent what he identified as an SOS to Hoover in Washington, begging the immediate aid of "assistants of yours . . . legal and otherwise, who would be beyond price in this emergency." As a PS at the bottom of the note, he added, "You will doubtless remember that we turned over to your organization the American Editors Syndicate, which I understand you have used very profitably in various foreign countries." The Syndicate was an office on Fifth Avenue that Macfadden had loaned to Hoover in the early days of the war, to use as a front in running spies to South America. After more than two weeks had passed, the FBI chief replied that since Macfadden's legal issues were not of the federal sort, there was really nothing the bureau could do for him.

Eleanor Roosevelt politely, if succinctly, continued to answer her former business partner's torrent of health-related letters, but other Washington connections withered after Franklin Roosevelt's death in 1945. Harry Truman, in particular, wanted nothing to do with his fellow Missouri native. When Macfadden wrote offering to publish a *Physical Culture* story praising Truman's fitness for the 1948 election, much as he'd done for Roosevelt in 1932, the White House press secretary scribbled "IGNORE" on the editor's note and filed it in clerical Siberia.

It was perhaps with this waning influence in mind that Macfadden embarked on what may have been the strangest project of his extraordinary career. In 1944, he filed papers in Florida to start his own religion. He named this new creed Cosmotarianism, which breaks down etymologically as "universal belief system," but in Macfadden's vision meant a marriage of equals, Christianity and physcultopathy.

Cosmotarianism's aim was certainly ambitious: to knead the world's great Christian religions into a harmonious nut loaf of faiths, with Macfadden's own exercise and diet wisdom as the binding agent. "Cosmotarian science," Macfadden wrote in the 1946 *Cosmotarian Seminary and Institute Catalogue*, "is interested not solely in man's soul: it teaches that health of mind and body are equally important with spiritual health; and provides physical culture, as well as metaphysical education in order to make man fit and worthy of his Maker." Jesus had gotten things off to a fine start, Macfadden conceded, but feeding the poor and tending to the sick wasn't sufficient in the modern age; the nutritionally downtrodden needed to be taught how to avoid defiling their blood with impurities. "Sickness is the result of sinning against the divine laws which the Creator provided for the body," Macfadden explained.

In the lessons Macfadden prepared, the Lord was referred to as "The One Divine Physician," a reminder to the Cosmotarian faithful to resist the diabolical promises of medical doctors. Macfadden stressed that Cosmotarianism, though it shared Christian Science's suspicion of modern medicine (as well as some of its favorite bread recipes), bore "no relation" to Ellen White's wholesome religion. Christian Scientists denied the efficacy of medical intervention because they saw humans as spiritual, not corporeal, beings. Macfadden, on the other hand, took literally— and quoted liberally—the words of St. Paul: "Know ye not your body is the temple of God?"

What inspired Macfadden to attempt to elevate his beliefs to divine status? Surely not piety. Macfadden had shown little prior interest in organized religion, except to identify himself as a Unitarian while running for the Senate. Were Macfadden blessed with a keener sense of irony, the reading materials created for Cosmotarianism could have been read as elaborate self-parodies of his own dogma. For novitiates, he prepared a ten-week course that cherry-picked quotes and stories from the Bible to underscore tenets of physical culture. One week's lesson stressed the importance of fasting. Questions to be answered and mailed in for credit toward enlightenment were listed at the end of each lesson, such as these from lesson 5:

1. Where in Exodus is fasting first referred to?
2. State passage in Exodus which refers to forty-day fast of Moses.

3. On what occasion did David fast for one day?

Below these, aspirants were expected to report on their efforts to incorporate the liturgy into their everyday lives:

I went without breakfast _____ times this week.
I undertook a twenty-four hour fast with nothing but _____ (State whether water or fruit juice.)

Potential converts were invited to study at three levels. The ten-week elementary course was for those "who are in search of Super-Health, Strength and Happiness." The Advanced Home-Study Cosmotarian Bible Course was aimed at "those who seek to enter this great new movement as organizers, teachers, lecturers, and counselors." These seekers were expected to familiarize themselves with a handful of sacred Cosmotarian texts, including the Bible, *The Encyclopedia of Physical Culture*, *Eating for Powerful Health,* and *Woman's Sex Life.* The highest level, the Professional Course in Cosmotarian Science, was specially designed for the chosen few who heard the calling "to enter into the Cosmotarian Ministry of God as practicing-ministers." Cosmotarian ministers would be trained in the healing of what Macfadden dismissed as "so-called ailments," including diarrhea, poison ivy, cancer, kleptomania, scurvy, ringworm, and malaria.

In May 1945, the spry, snowy-haired Macfadden took the stage in front of two thousand spectators at Carnegie Hall for "the Inauguration Services of the First Cosmotarian Church in New York City." Dressed in a white suit and white shoes, he delivered an hour-and-a-half-long speech, the first in a series of six talks on "religion through happiness." He preached his standard litany of dietary commandments ("beauty must be associated with a good digestion," he advised the female members of his flock), did push-ups, and stood on his head.

As a chartered entity, the Cosmotarian Fellowship Inc. continued for the better part of a decade without toppling the world's great religions. The name physcultopathy had never really caught on—even Macfadden rarely used it after the 1920s—so it's likely that Cosmotarianism was a failed last attempt to market his life's teachings to a mass audience. Macfadden might also have viewed the whole enterprise as a tax dodge, a way to funnel money from the foundation to cover his personal expenses. The meeting notes of the Cosmotarian Board of Trustees in 1952 contain this passage: "the Founder-Leader stated that during the past decade well over Two Hundred Thousand Dollars of capital funds were spent to bring COSMOTARIAN SCIENCE into the homes and lives of individuals throughout the nation." If Cosmotarianism was an expensive publicity stunt, the payoff was extremely poor. Little evidence of the religion seems to exist other than small press mentions of his first Carnegie Hall sermon. It is also possible that like many formerly influential men who resignedly enter their winter years, Macfadden approached his eighties wondering what his legacy would be. Buried in the Cosmotarian church's original by-laws was a section labeled "Article VII: National Shrine," stating that monies should be raised

for the purpose either of buying or leasing the Physical Culture Hotel at Dansville New York, and with the object of making this a living monument to Bernarr Macfadden, as well as an international shrine where his teachings, which he so courageously exposed [sic] and pioneered, shall be continued permanently to be taught and practiced as originally taught by him. . . .

Coincidentally, Fulton Oursler had also found religion, an awakening that had far greater ramifications than that of his boss. Upon Macfadden's departure from the company, Oursler had mothballed his teletype and returned to the New York City office, only to resign from Macfadden Publications within a year. Around the same time, he took up Roman Catholicism, with a passion he'd once reserved for exposing crystal-ball frauds and studying tarot cards. In 1949 he published *The Greatest Story*

Ever Told, a novelization of the life of Jesus Christ. It became one of the most popular books of the twentieth century, selling more than 4 million copies. Oursler died of a heart attack in 1952, two years before Twentieth Century Fox paid $2 million for the movie rights.

After eleven years of estrangement and bickering over money, Macfadden filed for divorce from Mary yet again in 1943, in Florida. The case was heard before a special master, an official who made recommendations to a judge. The proceedings were ugly. Most of the Macfadden children, perhaps mindful of the revocable trust funds their father had established for them, testified on his behalf. Basketfuls of the Physical Culture Family's dirty laundry were aired, from Bernarr's dipping Byrnece in cold water as an infant to Mary's chucking the razor box at her husband's mouth. Mary mocked her husband's "crackpot" health ideas and once again claimed that she'd invented *True Story*. She introduced as evidence a 1914 photograph of her inverted husband balancing on his cranium. "This was made when we first met," she told the court. "He probably would still be standing on his head except for me. I put him on his feet."

The special master and judge recommended a divorce, the sooner the better. Mary appealed the decision. Finally, in January 1946, the Florida Supreme Court affirmed the divorce decree of the lower court. The thirty-three-year marriage was over, though Mary would continue to contest the divorce until, and even after, her husband's death.

The trust funds for the seven surviving Macfadden children, all now of adult age, had been set up in 1931. Each started with $100,000 and had been arranged through the foundation. The mostly mundane correspondence that survives from Macfadden's final years consists largely of requests from his children to borrow thousands of dollars against these trusts. Macfadden criticized his offspring for their fiscal imprudence, but nearly always cut a check. In the end taking the loans turned out to be a sound financial strategy, since the foundation's finances finally got so shaky that the trusts were recalled.

Macfadden pitied himself when his children turned out to be average—"It is generally accepted that great men's sons rarely amount to anything," he whined in a note to Mary in 1949—but maintained mostly cordial relations with them. Half a century later, Brewster Macfadden and his wife, Peg, recalled Thanksgiving dinners in New Jersey at which the Father of Physical Culture presided over a healthy meal in the dining room while the younger generation snuck down to the basement to drink and smoke. By that time, the requests for paternal loans had dwindled to the occasional $50 or $100. Even these became impossible to fulfill eventually, as Macfadden was living on a fixed $24,000 annuity he'd purchased as an afterthought during the go-go twenties, and was charging whatever he could to the foundation. More than half of his income went to Mary's alimony.

The new *Physical Culture* arrived in 1943 at a moment when coffee, sugar, and many types of meat were being rationed for the war effort. Magazine buyers didn't need a martinet telling them what to cut out of their diets. Scientific medicine, meanwhile, was surging through a triumphant chemotherapeutic revolution. In the twenty years following the introduction of the antibiotic sulfanilamide, in 1937, an American's life expectancy from birth jumped from sixty years to seventy years. Killers such as strep throat, staph infections, and pneumonia were largely defanged, and deaths from childhood diseases fell 90 percent in a single generation. This overwhelming evidence refuting Macfadden's theory of blood impurity as the cause of disease certainly didn't help the relaunch of his magazine. The idea of healing through natural methods once again seemed like a quaint nineteenth-century relic. After all, why would a syphilis sufferer live on water alone for ten to fourteen days, followed by six to eight weeks of nothing but raw milk—as Macfadden prescribed to cure the venereal disease—when a doctor could inject him with a hypodermic needle of penicillin instead? *Physical Culture* limped along for another decade under at least four different names, its circulation and influence fading so much that the date of its final issue is unknown.

Macfadden never outgrew his orphan's love of attention. If anything, his needs only intensified as he grew older. In 1949 he parachuted from an airplane over Dansville to celebrate his eighty-first birthday. The publicity stunt was so successful that he repeated it over the Hudson River in 1951, the afternoon of his appearance on *What's My Line?* That jump was so popular that he scheduled another one in 1952, when a thousand Paris policemen blocked off a half-mile-long stretch along the Seine, awaiting Macfadden's arrival from the heavens. He leaped from nine hundred feet and alighted safely in an empty lot, some blocks away from the target site and an expectant crowd of several thousand Frenchmen. (His impact was cushioned by the special three-inch-thick foam-rubber soles of his shoes.) The *Chicago Tribune* reported that spectators disentangled Macfadden from his parachute and conveyed him to the riverfront, where a police boat collected him for a procession down the Seine as throngs on the banks yelled "Bravo, Macfadden!"

With his marriage to Mary officially over, Macfadden had captured headlines by taking a new bride. Johnnie Lee McKinney was a forty-four-year-old health lecturer and former interior designer when she met the seventy-nine-year-old publisher—a sturdy, vivacious blond Texas beauty molded in Macfadden's preferred silhouette. Theirs was a whirlwind courtship. He attended one of her talks in Manhattan, then hounded her into a lunch date at the New York Athletic Club. After a wholesome meal, the two proceeded to Johnnie Lee's apartment, where the vigorously amorous Macfadden demonstrated his usual distaste for small talk by unzipping his trousers to reveal what Johnnie

Lee called "the most exquisite sex organ I had ever seen on a man."* Johnnie Lee declined her date's unspoken offer—as well as his shouted proposal to marry her immediately—and ushered Macfadden into the hall. The next day Macfadden returned bearing armfuls of romantic gifts: raw nuts, dried fruits, and soybeans.

Macfadden whisked Johnnie Lee off to the Deauville, insisting that her health lectures were desperately needed by his clientele. She was surprised to find that his Miami office was staffed by three buxom secretaries whose idea of casual business wear was low-cut bikinis. Within the month, the suitor's ardency had worn down Johnnie Lee's defenses. She agreed to marry him. Five hundred guests attended their wedding at the Miami Beach Community Church on April 23, 1948. At the reception, the happy couple cut a whole-wheat cake. The next day, after a nude swim, Macfadden surprised his wife by mentioning for the first time that he was running for the governorship of Florida. Once again he flew around the state, promising to bring to life Ponce de León's vision of a Fountain of Youth. He finished seventh in a field of nine candidates, drawing fewer than five thousand votes.

After his election loss, the couple returned to New York City, where Macfadden explained to his wife that they'd be maintaining separate residences. Macfadden kept a small apartment once described by a visitor as similar to "the cell of an especially negligent monk." For Johnnie Lee, life with Macfadden was turning out to be one surprise after another. One Sunday night while the couple was listening to the radio with friends, Walter Winchell came on with the news flash that Johnnie Lee Macfadden was pregnant. She wasn't; her husband had leaked the story. Another time she confronted a man who'd been following her. He was a detective Macfadden had hired.

---

*For reasons about which this author prefers not to speculate, Johnnie Lee doodled a large heart next to this passage in her annotated copy of Robert Ernst's biography.

In early 1950, the Macfaddens began splitting their time between New York and San Bernardino in Southern California. Macfadden was a popular lecturer in the Los Angeles area, where California's first wave of health fanatics and bodybuilders had latched onto his message. (Two years earlier he had presented a trophy to Abbye "Pudgy" Stockton, naming her the Queen of Santa Monica's new myological Mecca, Muscle Beach.) Using foundation funds, Macfadden negotiated to purchase operating rights to the Arrowhead Springs Hotel from hospitality baron Conrad Hilton. This was to be the flagship property of a new chain of southern California health resorts that Macfadden had planned. He invested $100,000 in the facilities, only to see the Arrowhead go bankrupt a year later. Some time during this period, the fourth Mrs. Macfadden made an unscheduled visit to her husband's apartment and encountered him performing the marital act with another woman. They split, reconciled, parted again, reunited. When Johnnie Lee refused to join him in his parachute jump over the Seine—he'd insisted that she wear white tights with the letters M-A-C-F-A-D-D-E-N sewn onto the seat—Macfadden stormed off to Paris alone. Johnnie Lee sought a separation and support.

Right about then, Mary Macfadden published *Dumbbells and Carrot Strips: The Story of Bernarr Macfadden*, a demythologizing memoir of her marriage, written with the *Graphic*'s Emile Gauvreau. Mary wasn't satisfied with knocking her ex-husband off his pedestal; she wanted to get in a few kicks to the kidneys once she had him on the floor. All the scandalous secrets revealed during their divorce were now immortalized in print, supplemented with juicy new details. Macfadden was portrayed as a bumbling, barely literate hayseed whose savvy wife had dragged him to greatness, only to watch helplessly as he'd squandered their fortune with the help of his evil sycophant, Fulton Oursler.

Dumbbells was unfair and unbalanced, and some of it was untrue, but every page of it was immensely entertaining. "Reader comment upon *Dumbbells and Carrot Strips* is not likely to be feeble," wrote the *New York Times*'s reviewer. "Hilarious, fantastic, uninhibited and merciless are some words which come to mind." Johnnie Lee said Macfadden cried after reading the book. The tears may have shown a new level of self-awareness, for *Dumbbells* destroyed Macfadden's reputation as the elder statesman of fitness. When Macfadden tried to run for mayor of New York City in 1953 as the candidate of the Honesty Party, the board of elections refused to validate his petitions. His name never made it onto the ballot.

In 1954, the Utah Aeronautics Commission denied Macfadden's request to stage an eighty-sixth birthday leap into the Great Salt Lake. Judging from photographs taken around this time, they did him a favor. Though he'd long estimated his own life span at 125 to 150 years, Macfadden was old, and he looked it. His proud physique had withered, and he'd taken on the slight stoop of the elderly. He was helpless when Berwyn barged into his office in mid-October, furious, and physically attacked his father, whom he believed had phoned the dance instruction kingpin Arthur Murray to get Berwyn fired from his teaching job at Murray's studio. Macfadden pressed battery charges against his son. Pictures appeared in newspapers of the shrunken former strongman sporting an ugly black eye. The exhausted look on Macfadden's face was understandable, since he was now being pursued by the IRS for $18,000 in unpaid taxes on alimony monies. Two sets of divorce lawyers were hounding him as well. Mary was angling to nullify their divorce—in a vain hope that she might get her money back from the Macfadden Foundation—and Johnnie Lee was trying to arrange one, "on grounds that he subjected her to cruel and unusual treatment," according to the *Los Angeles Times*. Macfadden spent months trying to track down Johnnie Lee's first husband in Texas, hoping to prove that she was a bigamist (and therefore not entitled to support). He failed. Disbursements to both women left him with only $100 a month to live on, and eventually he fell behind on payments.

The New York Supreme Court judged Macfadden in contempt

for failing to pay his debts in late December 1954. "I must raise $10,000 in ten days or go to jail," he told newsmen on Christmas Day. The foundation was legally barred from lending money, even to its founder, and since Macfadden no longer owned anything of value to offer as collateral, taking out a loan was impossible. He asked anyone who wished to lend him money to send it in care of his friend, former heavyweight champion Jack Dempsey. "Please help me," he pleaded, "this is my worst Christmas in 86 years."

Macfadden dodged jail temporarily when a judge postponed his incarceration. He used this break to slip into Canada, where he camped out at the Sheraton-Brock Hotel in Niagara Falls, Ontario, and made occasional clandestine visits to Dansville. In a March conversation with a local reporter, Macfadden expressed bitterness at his fate. "People today do not want to be helped and if I had it to do over again I would have kept the money," he said. "I have done a great deal in trying to make people realize they can have a full healthy life by following the practices of physical culture, but they would rather suffer from sickness and die when they are too young than do anything which will cut into their pleasure."

After a few months, he moved into a room at the Earle Hotel in Jersey City, New Jersey, in hopes of eluding the long arm of New York State law. The Hudson County sheriff arrested him there on April 18, 1955. A small story in the *New York Times* noted that Macfadden "faced his second night in jail cheerfully today, praising the food but lamenting that the cell he shares with two trusties is too small for exercise." On his third day behind bars, he raised the $10,000 bail and was released. His unshakable confidence had been shattered; the foundation's press agent, Ed Bodin, later said, "He went to pieces after that jail thing . . . something broke in Macfadden. I saw it in his eyes." One night in June, a fellow resident at the Earle tried to commit suicide by inhaling gas and ignited an explosion that blew doors off their hinges and cracked plaster. Macfadden slept through the blast.

On the evening of October 7, the manager of the Earle found Macfadden unconscious in his room. Macfadden had complained of two recent bouts of biliousness and had evidently sought to cure this third outbreak by fasting. Instead, he'd starved himself into a serious case of jaundice and dehydration. Taken to the emergency room at the Jersey City Medical Center, the enemy of organized medicine was "fed intravenously, catheterized, x-rayed and examined head to foot" by physicians. Doctors suspected a duct stone, and discussed (out of earshot, one hopes) exploratory surgery, but decided against invasive measures based on Macfadden's poor condition. On October 12, New York City radio stations broadcast the news that Bernarr Macfadden was slipping away. At 10:25 p.m. he was pronounced dead at the age of eighty-seven.

His last act as a physical culturist, it was later revealed, had been to mail a letter to President Dwight Eisenhower, who'd just suffered a heart attack. Macfadden had suggested some exercises to strengthen the president's cardiac muscle.

For the final decade of his life, Bernarr Macfadden protested long and hard that after starting the foundation he was no longer a wealthy man. No one wanted to believe him. In his final will, Macfadden left $1,000 to each of nine relatives and ordered the remainder of his estate to be split equally between his children and the foundation. He left nothing to any of his wives. Not that it mattered—Macfadden had died virtually penniless. No one inherited anything.

Not surprisingly for a man who shrouded so much of his life in secrecy, rumors began to circulate shortly after Macfadden's death that he had buried large caches of money—Johnnie Lee claimed that he had mentioned amounts up to $4 million—in locations around the country. The most prominent legend told of Macfadden sneaking away from the hotel at Dansville with a large metal box and a shovel, presumably to inter a chunk of his fortune in the earth for later retrieval. The year after his death, the Dansville paper noted that the new owner of the Physical Culture Hotel "has bought 100 shovels, two surplus mine detectors and two machines called M-Scopes used to find old buried pipelines" in order to exhume the loot. Nothing was ever uncovered. A bulldozer operator did unearth $89,000 in cash on Long Island in 1960 not far from where Macfadden had once operated a health home. Johnnie Lee laid claim to the loot but a judge dismissed her petition. Legends of buried treasure evidently die hard. Though the Finger Lakes region of upstate New York has never been known for deposits of precious ore, a sign in the window of the Dansville Public Library strictly forbids metal detecting on its property.

Mary Macfadden spent a decade successfully fighting the Bernarr Macfadden Foundation's efforts to evict her from the Englewood home. The mansion fell into disrepair, its once-beautiful grounds overgrown with weeds. Brewster said it reminded him of Miss Havisham's house in *Great Expectations*. Mary closed off most of the home's rooms, lost her sight, and was cared for in her final years by an Irish couple. She died in 1969. Today much of the original estate is a public park.

Susie Wood spent her retirement years living north of New York City. She occasionally received visits from Helen Macfadden Wiegers and her children, Dan and Joan. The two eventually figured out that the woman they were instructed to address as Aunt Susie was actually their maternal grandmother but never acknowledged it in her presence. "There was always a lot of intrigue on the Macfadden side of the family," says Dan Wiegers.

Johnnie Lee spent the next three decades lecturing on health (she was a guest on *The Tonight Show with Jack Paar*) and advising private clients. She, too, published a memoir of her years with Macfadden, the airbrushed remembrance *Barefoot in Eden*. Johnnie Lee spent her final years in New York City. She died in 1992.

Macfadden's children, as he feared, grew to be much like other members of their generation: They fought in wars and bore many grandchildren; some lived happy suburban lives, others suffered from serious alcohol and emotional problems. At the age of eighty-two, Brewster, one of Bernarr Macfadden's two surviving children (Berwyn has been institutionalized in New York for many years), is a shocking sight—a funhouse mirror twin of his father, a six-footer whose frame is packed with an additional fifty pounds of muscle. His firm handshake could have brought Sandow to his knees. The father that he remembered from the 1950s was strict, but loving, too: a physical-culture paterfamilias who could turn a blind eye to his adult children enjoying a cocktail and a cigarette.

The Bernarr Macfadden Foundation did not save America from its sad fate as a nation of flabby mollycoddles. Instead, the foundation expired slowly, its various properties sold off to satisfy creditors and lawsuits in the generation following its founder's death. It ceased to exist in 1978.

But that is not the end of the story.

EPILOGUE

The Physical Culture Hotel still stands, barely, on a hillside above Dansville, New York. Its redbrick shell is visible from Highway 390 once the leaves have fallen, so should you find yourself driving from Elmira to Rochester on a late autumn day, tap the brakes and look to the right to get a peek at the building that, between the tenancies of James Caleb Jackson and Bernarr Macfadden (and one very fruitful visit from Seventh Day Adventist founder Ellen White), can lay claim to the title of Holiest Shrine in American Alternative Health. If you reach the Tops Supermarket without spotting Macfadden's cathedral, turn around—you've passed it.

I put off making my first pilgrimage to the P.C. Hotel until this book was nearly complete. Having spent four years poring over photos from its glory days, I was shocked at how badly it had aged. The building has changed hands a few times since Macfadden's death, and hasn't been used at all since 1971—roughly the moment America's smoldering interest in health food, exercise, and preventive medicine reignited. Rolling green lawns where health seekers once lounged and frolicked are now plugged with mature trees. The ornate wooden porch collapsed years ago, and when I arrived for my prearranged visit—only invited guests are welcome these days, since the building is on private property—Dansville Mayor Bill Dixon had to jimmy open a boarded-up doorway to gain entry. Inside, some skeletal evidence of the golden age of physcultopathy remained, such as the handsome iron staircases and classical columns that frame the entryway, and a lone personal sauna in what was the basement hygienic bathing room. But thirty years of neglect and teenage rebellion had taken their toll. Every window had been smashed. Graffiti covered the walls of the formerly elegant dining room. The physical-culture kitchen where millions of juice oranges once met their fates basked roofless in Macfadden's beloved sunlight.

Still, Dansville is surely the only town in America where most of the residents recognize the name Bernarr Macfadden. Though Macfadden was arguably the most influential figure in the history of this nation's love-hate relationship with exercise, no one outside of upstate New York acknowledges him nowadays, except for bodybuilding historians and the Polar Bear Club, those hypothermia enthusiasts who splash around in icy water every New Year's Day to the delight of local news crews. They claim Macfadden founded the club in 1903, which is possible; I couldn't find any evidence that he'd done so.

Trying to trace particular preventive health theories to an original source is tricky, since most can be summarized in four words that haven't changed much since they were first uttered on the streets of ancient Athens: Eat less and exercise. Macfadden's great achievement was to take the oddball health ideas of a bunch of bushy-bearded nineteenth-century abolitionists and freethinkers and repackage them in a form that was palatable for a mass-media audience that didn't exist until after his death. Many of his causes that seemed shocking in the early part of the twentieth century—sex education, eliminating white sugar and bleached flour from one's diet, strength training for women, rehabilitating heart-attack victims with aerobic exercise—are now taken for granted.

Macfadden had definitely been on to something when he planned to start a chain of spas out west, based on the model of the Physical Culture Hotel. It was the right idea at the wrong time. In the years after World War II, the epicenter of America's diet and fitness world migrated from Boston, Battle Creek, and Dansville to California, drawing those who sought to reinvent themselves. Many health crusaders who succeeded on the Pacific coast were pushing ideas similar to Macfadden's. Jack LaLanne, protégé of Paul Bragg, became television's first fitness celebrity. He even started performing stunts to show off his vitality—swimming to San Francisco from Alcatraz shackled in handcuffs just a few months before Macfadden's death in 1955—and would do so, usually to mark a birthday, into his eighties. Nutritionist Gayelord Hauser, who'd written about the future of eating for *Physical Culture* and went on to advise Hollywood stars such as Marlene Dietrich and Gloria Swanson, had the number-one book in America in 1951, *Look Younger, Live Longer*. His dietary prescription for yogurt, wheat germ, and barrels of vitamins made *Look Younger* a popular item in America's skyrocketing number of health-food stores, where half its copies were sold. Los Angeles nutritionist Adelle Davis published her multimillion-selling masterwork, *Let's Eat Right to Keep Fit* in 1954, just in time to introduce home-baked Graham bread and milk guzzling to the Baby Boom generation. If any municipality on earth could have been said to resemble Macfadden's original utopian plan for Physical Culture City, it might have been Berkeley, California.

The greatest health publisher of the 1950s and '60s, Pennsylvania's Jerome I. Rodale—who first applied the word *organic* to agriculture—was a fervent admirer of Macfadden and a longtime *Physical Culture* reader; he'd used physcultopathic methods to cure himself of ailments as a sickly child and gone on to found a publication in the 1930s called *True Health Stories*. Building on the chemical-free farming techniques of Britain's Sir Albert Howard, Rodale started a magazine in 1942 for those who wanted to eat pure food, called *Organic Gardening and Farming*. In 1950, Rodale launched the alternative health bible *Prevention*, which looked like a less muscular, more holistic *Physical Culture*, updated from the era of neurasthenia to the nervous atomic age. By the early seventies it sold a million copies each month.

The AMA predictably labeled Rodale a quack. The counterculture revered him as a sage. He died of a heart attack while taping an episode of *The Dick Cavett Show*, and his son Robert took over the company, going on to build huge markets for magazines such as *Runner's World* and *Men's Health*. Those publications and others like them (especially Joe Weider's *Muscle and Fitness*, required reading for the protein-shake crowd) that trod paths blazed by *Physical Culture* helped spur the growth of a multibillion-dollar fitness industry. If the first two pillars of Macfadden's original vitality formula—fasting, hydropathy, exercise—have not exactly taken root, the third has triumphed to a degree that the man himself couldn't have imagined in 1955. Even the majority of Americans who don't exercise regularly are expected to feel guilty about not doing so.

The health gurus whose influence mushroomed in the 1960s, J. I. Rodale and Adelle Davis especially, were the beneficiaries of what have become periodic backlashes against processed foods and chemotherapeutic medicine. In 1959, much of the nation's pre-Thanksgiving cranberry crop was found to contain a potentially cancer-causing herbicide. The following year, mothers who had taken the drug thalidomide as a sedative began giving birth to babies with grotesquely malformed limbs. On the heels of those scares came Rachel Carson's book *Silent Spring*, which demonized pesticides (DDT, most famously) and ushered in a new wave of nature consciousness. The rapid acceptance in the mid-seventies of holistic medicine and farmer's markets, and a renewed reverence for *mens sana in corpore sano*, would not have been possible without the proselytizing on a mass scale that Macfadden did in the 1930s.

Macfadden's universal method of healing—fasting—remains largely untested. He would probably argue that pharmaceutical companies have suppressed a therapy that is 100 percent natural and free. It is also true that fasting is unpleasant and counterintuitive—parents who cut off their children's solid food for three weeks, as Macfadden did when his kids came down with whooping cough, tend to end up on the evening news exiting a courthouse with their coats pulled over their heads. (One could also argue that the only significant difference between fasting and anorexia nervosa is the degree of control a person maintains over deciding when to start eating again.) Yet studies have shown that arthritis, among other maladies, can be alleviated by fasting. Research conducted by the National Institute on Aging on rats and mice showed that intermittent fasts such as Macfadden's food-free Mondays might help protect the body against the effects of Huntington's disease, Alzheimer's, Parkinson's, and stroke. One fascinating study indicated that cancer patients who fasted forty-eight hours before undergoing chemotherapy suffered fewer negative effects from the treatment; the body's healthy cells are believed to enter survival mode when food is cut off, while cancerous ones, oblivious to the warning signs of starvation, do not. The calorie-restriction community, probably the most extreme group of food faddists around today, is carrying out long-term experiments on itself that, if successful, will prove a theory that Macfadden had put forth in *Physical Culture* in 1926, eight years before the famous Cornell University study that showed a drastic decrease in calories might greatly improve life span.

The battle between regular and irregular practitioners has largely subsided. Osteopaths and chiropractors are covered by Medicare and receive referrals from physicians. (Macfadden would have stood on his head for joy had he lived to hear the 1987 ruling by a U.S. District Court that the AMA and other allopathic trade groups from his despised "medical monopoly" had violated antitrust law by discriminating against chiropractors.) Practitioners of naturopathy, physcultopathy's near-identical cousin, are now licensed in thirteen states, an acceptance that was speeded along by naturopaths conceding that the germ theory might have some weight behind it after all. The detoxification methods that Macfadden, Kellogg, and others popularized at their sanitariums have made an amazing comeback in recent years, as has in-home natural childbirth—a development with which Mary Macfadden might take issue.

The "natural diet" is more popular than ever, especially among the Hollywood stars Macfadden so carefully cultivated in *Physical Culture*. Excise the chitchat and magical thinking from diet books such as *Fit for Life* and *The Zone* and the *Skinny Bitch* series, and the advice they contain could have come straight from the 1912 *Encyclopedia of Physical Culture*. (As could the mantra of Michael Pollan's more cerebral *In Defense of Food*: "Eat food. Not too much. Mostly plants.") If one were to update the batting averages of Macfadden's crusades, the numbers might look like this:

Against prudishness	1.000
Against medicine	.333
Against corsets	1.000
Against muscular inactivity	.600
Against alcohol	.250
Against cigarettes	.750
Against white bread	1.000

This would raise his batting average to an imposing .705, surely high enough to earn a place of honor in the health reformer Hall of Fame. There's still a chance that Dansville will be remembered as his Cooperstown. As I was finishing this book, the town received a $2.5 million grant to help underwrite the latest owner's plan to refurbish the Castle as a medical facility. He even opened up the grounds to a treasure-hunting club armed with metal detectors, but thus far, the legendary Macfadden cache has remained elusive.

Some of Macfadden's most lasting influence—much of it bad—springs from his work as a journalist. The *Graphic*, both in dragging other newspapers down to its level and in begetting Walter Winchell and Robert Harrison, the founder of *Confidential*, incubated the American gossip industry and abetted the rise of celebrity culture. Macfadden may be the spiritual grandfather of Whole Foods and the Canyon Ranch spas, but his offspring also include *Us Weekly*, the E! Network and TMZ.com. The *National Enquirer* and its tabloid kin are direct descendants of the *Graphic*, as is the *New York Post*, with its irresistible combination of cheesecake photos, crime sagas, and, of course, unrelenting celebrity coverage. Macfadden published a lot of skin in his magazines and newspapers, and paid a lot of lawyers to argue that he had the right to do so. In the process he undoubtedly made life a lot easier for Hugh Hefner, who was laying out the first issue of *Playboy* on his kitchen table in Chicago around the time Macfadden was plummeting toward the Seine.

The spirit of *True Story* hovers over each weepy Barbara Walters special and every issue of *People* magazine in which an actress declares "I want a baby"; it haunts the worst precincts of daytime television. The *Jerry Springer Show* period of the 1990s, when America's top growth industry seemed to be producing mothers who slept with their daughters' boyfriends, owed the same debt to Macfadden that the bands of the British Invasion did to bluesman Robert Johnson. (The stalwart *America's Most Wanted* owes the same debt to *True Detective Mysteries* that an Elvis impersonator owes to Elvis Presley.) In many ways, Oprah Winfrey has built herself into the brand that Macfadden always wished to be: The popular figure behind a multimedia empire, who dispenses moral advice to millions; a political force whose ability to sway voters is taken seriously all the way up to the White House; and, both through her own best-selling diet and exercise books and by promoting works such as the *You: An Owner's Manual* series, this century's number-one driver of interest in physical fitness.

With the death of *True Detective* a few years back, *True Story* remains as the sole newsstand survivor among Macfadden's commercial magazines (though the arts journal *The Dance* began life as his *Dance Lovers*). Now almost ninety, *True Story* still serves up the same menu of mildly salacious confessions that it always has, if on a more modest scale. Its most conspicuous print progeny aren't magazines at all, but the books in the preposterously popular *Chicken Soup for the Soul* series: dozens of editions, over 100 million copies sold, each composed entirely of nonfiction parables submitted by readers. If there were any justice in the world of publishing, the *Chicken Soup* folks would lay a wreath on Macfadden's grave every August 16 to honor the birthday of the man who taught America to spill its secrets.

But first they'd have to find his burial site, which isn't easy. Tucked away in Woodlawn Cemetery in the Bronx, Macfadden's surprisingly small tomb marker is the inverse of his Physical Culture Hotel: obscure from a distance but unmistakable as one draws near. It's the only one within a hundred yards ornamented with a bust of its occupant—clothed, surprisingly, in a natty bowtie. Mary chose to rejoin him for eternity, though as in life she's been relegated to a spot in her husband's shadow, on the ground in front of him alongside three of their children.

Engraved on his podium are the only words by which he wished to be remembered:

BERNARR MACFADDEN 1868–1955
FATHER OF PHYSICAL CULTURE

For better or for worse, we are all his children.

My Life on Physcultopathy

Like Louis Pasteur and Dr. Jekyll, Bernarr Macfadden was a great believer in testing his new theories on himself. Thus it only seemed fair that to more fully understand the ideas of a man who devoted his life and fortune to physcultopathy that I, too, should play the test subject. This wasn't easy. For all the millions of words he published on the subject of health, and despite having personally overseen regimens for thousands of visitors to his sanitariums and health homes, Macfadden never bothered to boil his wisdom down to a series of simple steps to be followed.

Macfadden was, however, a great lover of efficiency, so I don't think he would mind if I synthesized some of his core principles. With a nod to Macfadden's gift for hyperbole, I'll call this list:

BERNARR MACFADDEN'S PHYSICAL CULTURE DIET
1. PREPARE TO FEEL HUNGRY.
2. ELIMINATE CAFFEINE AND ALCOHOL.
3. START WITH A FAST.

4. CUT YOUR FOOD INTAKE IN HALF.

5. EAT RAW, FIBROUS, AND BLAND.

6. FLETCHERIZE, OR CHEW CHEW CHEW.

7. EMBRACE THE MIRACLE OF MILK.

8. WALK TILL YOU DROP.

9. BUILD STRENGTH.

10. SHIFT TO TWO MEALS A DAY.

1. **Prepare to feel hungry.** Macfadden wrote: "The most important and essential principle of a reducing diet is that the total food-intake, as measured in calories, should be materially below the weight-maintaining requirement." Fat is the product of overeating; the cure is undereating. You cannot lose weight until you wrap your head around this fact. Try to maintain the attitude that if you are hungry, you are losing weight. Hunger is your friend.

2. **Eliminate coffee and alcohol from your diet.** These were two of Macfadden's trinity of everyday "poisons," along with nicotine. He felt that they poisoned the blood and dampened one's natural vitality. As a substitute, he suggested "the positive cultivated love of life." Should that prove insufficient, he also thought spending more time in the nude would help.

3. **Start a weight-reduction program by fasting for three days.** Fasting was one of Macfadden's keystone therapies. He believed that it could cure everything from epilepsy to asthma to cancer. He considered fasting to be a necessary detoxification, or "housecleaning" for the body. He recognized that fighting fat is first and foremost a psychological battle, and noted that would-be reducers who fasted were encouraged by their quick weight loss. Some, but not all of this initial slimming is due to water loss and the total evacuation of the alimentary canal.

4. **Cut your food intake by roughly half.** Not half of what you'd like to eat, but half of what your body needs to maintain its weight. (Any smaller reductions cause the body to "utilize its supplies with greater economy.") On average, he said, a moderately active adult needs about 2,500 calories a day. Instead of aiming for 1,250 calories, though, he recommended eating ten 100-calorie portions of low-fat foods, especially skim milk. Also consume as much as you want of leafy vegetables (spinach, lettuce), bran, tomatoes, pod vegetables (string beans, peppers), and root vegetables (other than potatoes and sweet potatoes). If this program sounds familiar, it should—Dr. Dean Ornish formulated a very similar plan in his million-selling *Eat More, Weigh Less*.

5. **Eat as many high-fiber foods as you can, cooked as little as possible, and with as little seasoning as possible.** Macfadden was obsessed with maintaining a clean colon, and since most people probably won't share his enthusiasm for enemas, eating a lot of fiber (or "cellulose," as he called it) is the best way to make this happen. Macfadden was a raw-food lover because he liked how uncooked foods fill up the stomachs of dieters; because one is less likely to eat them prepared with oil, butter, or cream; and because they help prevent constipation, which Macfadden—a man who cultivated the nickname B.M.—saw as the root of much health evil. Macfadden disdained spices because he wanted people to appreciate the taste of untreated foods.

6. **Fletcherize.** This term comes from the turn-of-the-century food faddist and weight-loss guru Horace Fletcher, aka the Great Masticator. This rule is simple. Chew your food twenty, thirty, forty, or however many times you need to make it slide down the back of your throat. (Fletcher once proudly pulverized a tough green onion with more than one hundred chews.) No swallowing should be necessary. The more the food is

chewed, the greater the bulk in the stomach, the less the sensation of hunger.

7. **Embrace the miracle of milk.** Macfadden considered milk to be the greatest of all foods and often prescribed milk-only diets for therapeutic reasons. Anyone on a reducing diet, he wrote, should consume a quart a day to provide calcium and protein; anyone trying to bulk up should consume several times that amount. I would guess that he would have approved of organic fortified soy milk for those who don't do dairy.

8. **Start walking.** Macfadden considered exercise to be the most important activity in a man or woman's life, with the possible exception of sex. He called walking "the best of exercises." His regimen was the same for all people: Walk until you are fatigued, rest, and then walk some more. It was, in his eyes, impossible to walk too much. Aim for at least three to five miles a day. Once you've built up some fitness, you can start running or biking. Or you can just keep walking (shoes are optional).

9. **Build strength.** Macfadden believed in "muscular activity"— meaning calisthenics or sports—six days a week. A few guidelines: Mornings are usually the best times for strength building. Work on perfecting your form rather than trying to lift the heaviest weight possible. Use a variety of exercises to keep from getting bored. Remember that strengthening the spine is the most important goal of any exercise program.

10. **Eat only two meals a day.** The hardest part of a diet is keeping the weight off. One of Macfadden's first and favorite discoveries, the two-meals-a-day plan, is a variation on the weight loss truism "eat when you're hungry." It consists of a late breakfast/early lunch (around eleven a.m. or so) and an earlyish dinner (at five or six). He found that, while people tended to eat a bit more at each of these meals than they did

at each breakfast, lunch, or dinner, the overall amount of food consumed was less, the equivalent of about two and a half meals under the three-meal system. "The two meal plan, because of its simplicity, is therefore the easiest way to discourage overeating," he wrote.

EXPERIMENT 1: MY LIFE ON THE P.C. DIET (BEGINNER VERSION)

Obeying all ten of Macfadden's laws from a cold start would be almost impossible, so I decided to ease into the regimen slowly. On Day One of my experiment, I weighed 189 pounds. My chest measured 39 inches, my waist 39 ½ inches, and my hips 41 inches. Five years ago, those numbers had been 41-35-40, and I had weighed 170 pounds. In my closet, I had a $5,000 custom-made suit cut to those exact measurements; I hadn't been able to wear it in years. Wearing that suit again was my goal.

I started with exercise. In the 1931 edition of his *Encyclopedia*, Macfadden provided a simple twelve-exercise dumbbell routine. The same weight can be used for each movement—he recommends one- to two-pound dumbbells, but because I felt ridiculous, I upped that to ten-pound dumbbells—and each exercise is done between five and twenty times or until exhaustion. Because Macfadden thought the back muscles were the most important ones, most of these dumbbell moves work the spine and midsection as much as the arms and legs. (If you've ever worked on your "core," it's the same thing.) I added three exercises for strengthening the lower back and one for the lower back and hips. I also chose four exercises for the abdominals, two for the neck, and four for the legs. I finished each workout with push-ups to exhaustion as a barometer of my strength; at the outset of my experiment, I could do eleven. My entire routine of twenty-seven exercises took half an hour to complete.

Although the terms *cardio* and *aerobics* hadn't been invented when Macfadden was still in the fitness business, they describe the benefits of what he felt were the two greatest exercises, walking and running. I upped my walking to two to three miles a day and started running two miles every other day.

The first couple of workouts were awkward, but as my muscles started to get the hang of exercising, I began to enjoy myself. After about three weeks, I could see some definition returning to my arms. The back exercises made my spine straighter, and I felt as if I'd grown an inch. If you've never worked with weights before, I recommend it if only to remember what it feels like to have the sensation that your muscles are all connected, that you can flex a muscle in your shoulder and feel it all the way through your back, butt, and into your legs. For me, this took about five weeks. During that time, I increased my running, adding a mile to my total distance each week. After seven not particularly grueling weeks, I was running eleven to twelve miles a week and feeling fairly fit.

I was not, however, looking any more fit. I had continued my usual diet of consuming whatever I wanted, including a couple drinks in the evening, maybe more on weekends. Almost fifty days into my new Macfadden lifestyle, I'd actually *gained* two pounds, bringing my weight to 191. These extra pounds were likely muscle that had been added from my strength workouts, but my pants weren't fitting any more loosely. It was time to get serious. It was time for a fast.

Macfadden outlines dozens of different types ·of fasts in his writings, but the water-only version was first among equals. You can't just jump into a fast—you have to be mentally and physically ready for it. I knew that I'd suffer from caffeine-withdrawal headaches, so I limited myself to one cup of coffee for two days before. The first day I made it until noon before my head began to feel like it was cracking open. The second day I made it until around 2:30. Both days I had a few sips of coffee in the afternoon and felt better. On the third day, the day before my fast, I sipped a little tea in the morning and afternoon and ate lightly, mostly fruit and vegetables. Part of the goal of the fast is to clean out the alimentary canal, so the less that's in there, the better.

The first of my three fasting days, a Friday, was strange, and not just because I had a mild, coffee-withdrawal throb in my forehead. It was only when I'd forsaken eating for a day that I realized just how much my schedule revolved around the buying, preparing, and consuming of food. When fasting, you have no such markers to break up your day. I started to feel a little hungry at eleven a.m., and the mild pangs came and went most of the day. I'd expected to feel exhausted from low blood sugar, especially in the midafternoon when I usually have to drink a pint of strong coffee to keep from falling asleep. But I put in a full day at work and had no trouble sleeping that night.

Saturday, I was hungry. Not starvation hungry, but hungry enough that I made a trip to the market and started scouting out the fruit that would be the next thing I would eat—in forty-eight hours. Macfadden recommends breaking the fast with a piece of juicy fruit and a glass of milk. After poking and prodding dozens of apples and peaches, I decided on three very large nectarines. (Macfadden hadn't specified the *size* of the fruit.) I passed the rest of the day reading and doing laundry. I had vivid visions of red wine and tomatoes throughout the day. I didn't have a coffee headache, but I did feel lightheaded. In the late afternoon, I tried to replace a part on my lawnmower but couldn't keep track of the sequence of steps I was supposed to follow. When I turned on the machine, the horrible grinding sound I heard told me that I'd skipped at least one. I went inside and passed out for a couple hours.

The third day of fasting is often described as the breakthrough day, when your body feels cleansed and your mind opens to the universe. I felt pretty much the same as I had the day before. The temperature hit 90 by eleven a.m., but I had to throw on a sweater because of chills. After a while, my hunger subsided. A sort of comfortable awareness that my body was empty replaced the pangs; the feeling was sort of like looking down with satisfaction at a sidewalk I'd just shoveled. Macfadden had warned that one's breath and body odor can become unbearable around this point in the fast. Aside from a horribly coated tongue, I didn't notice much difference. Something new was definitely happening inside me, though. I'd been drinking a glass of water every hour, but my urine continued to run the color of lager. Macfadden would have chalked this up to the exodus of impurities from my bloodstream. (My physician, when I sheepishly mentioned this later, chalked the change up to my body's entering ketosis, a possibly toxic state that also happens to be the physiological basis for the Atkins Diet. "That must have been *great* for your kidneys," he said.) Macfadden also says that one's sense of smell becomes heightened during a fast, and this was definitely the case with me. I picked up the nectarines and could smell the tart sweetness of the fruit lurking beneath the skin. I put them down before I did something I'd regret later.

Around lunchtime on Sunday, I walked a mile to the store and purchased the richest-looking milk I could find. (I opted for a half gallon of organic whole milk that cost about double what the regular milk did.) When I returned, curiosity got the better of me and I weighed myself: 183.8 pounds. I'd lost 7.2 pounds in two days, presumably mostly water. Macfadden was right about the confidence boost. The fast suddenly seemed like a brilliant idea, and I couldn't wait until Monday to see how much more I could lose. I took another long nap, watched some TV, and kept counting the hours until I could eat again. Before I went to bed, I washed all the dishes that had accumulated over the weekend: a single water glass.

Monday morning, fast-breaking day, felt a little bit like Christmas. When I walked down to the basement, the scale gave me my first present. My weight had dropped to 181.8, bringing my total loss to 9.2 pounds since Friday. For two days, I was allowed three tiny meals of one fresh fruit, a little dried fruit, and a glass of milk. I went into the kitchen, sliced the nectarine into twenty pieces and chewed each as if it were Kobe beef; I could taste every rivulet of juice as it trickled between my molars. I have never enjoyed fruit as much. The milk, measured out in a Pyrex cup to exactly eight ounces, was even better. Its rich, sweet, buttery texture and taste were like melted ice cream, and I savored each tiny sip, making the glass last for twenty minutes. Each nibble of the plump, chewy dried apricot that rounded out my meal unleashed an explosion of concentrated fruitiness.

Monday passed as a fairly normal day, and I didn't feel particularly hungry at any point. You really can't fit too much in your stomach when it's been idle for three days, anyway. At the end of the workday, I called a friend who's a personal trainer and mentioned that I was still feeling a little logy. He suggested I eat some salt and drink a little coffee in the morning. I knew that salt loss was largely responsible for my dropping nine pounds, and I wasn't entirely eager to add it back in; I was also feeling holier-than-thou as I watched my coworkers slug down their pots of coffee. When I cut up a tomato that night and sprinkled salt on it, however, the flavor was extraordinary, rich and full-bodied. I tried a slice with table salt and one with kosher salt and had no trouble distinguishing between the two flavors (the Morton's had a slightly chemical, processed taste). My milk aperitif was the perfect contrast.

I awoke Tuesday aware that something was different. I felt . . . amazing. I'd slept deeply, and my head was clear; a tiny cup of weak coffee only heightened the sharpness of my thoughts. I could only assume that this was the delayed arrival of the euphoria I'd been promised. I did a light strength workout and ran two and a half miles. At work I cleared my desk. Even with my salt consumption, my weight had crept back up to just 182.6.

Wednesday I still felt great and ran four miles. My weight held steady at 182.6. I ate some green vegetables and fish. I noticed that a months-old patch of acne on my forehead had vanished. That night on the train I ran into Bela, the mother of one of my son Alex's best friends. Bela grew up in India, where she'd fasted frequently with her family. "Let me guess, after you were done you didn't crave the same things anymore, right?" she asked. She was right. Whereas I usually ran home to have a glass of wine or three after a long day, I wasn't feeling that need. A week went by without alcohol, and I still felt no craving for it. I drank a little coffee, but was probably taking in about a third of what I had been.

Any weight-loss book will tell you that the danger of extreme calorie restriction is that your body will burn muscle along with fat, and that your weight will rebound quickly. To the former, all I can say is that I could do twenty-two push-ups before the fast and twenty-two push-ups after the fast. My chest measurement was unchanged. As for the latter, my weight seemed to hold steady over the next couple weeks, as I indulged in a few nice meals and a few drinks. My energy level, however, remained through the roof. I awoke every day at five feeling rested and immediately got to work or had a workout.

The Sunday before my fast, I had run 5 miles, the longest I'd gone in over a year. I had hoped to build up to half-marathon distance as I had in the past: by adding a mile a week to my longest run each week. The Sunday after the fast, I ran 5 miles again, which should have put me on target to run my 13.1 miles eight weeks later. I was feeling a surge of energy, though. The following Sunday I wanted to run 6 miles, and felt good enough to run 8. The next week I thought I'd try for 9 and did 11. On the fourth Sunday after the fast, I hoped to make 12, and did 13.5.

Clearly something had changed. The first time I'd run a half marathon, after two months of workouts with a professional trainer, I'd collapsed into a long afternoon nap, needed to gobble down handfuls of Tylenol to combat the muscle soreness, and spent the better part of the week walking stiff-legged. This time, I took nothing and felt fine Monday morning. Tuesday I ran three miles and lifted weights with no problems. In the week after the fast, my weight slowly crept back up to 186, but with my increased running I continued to lose a pound a week until, seven weeks after starting my fast, I was back down to 183. My strength had improved—I could now do thirty push-ups. And even with a return to some of my non-Macfadden-approved habits (alcohol, coffee, meat) I'd gained a half inch in my chest, lost two in my waist, and lost a half inch around my hips. The custom-made suit didn't fit perfectly. But it fit.

EXPERIMENT 2: MACFADDEN'S
"NATURAL FOOD" DIET

"Cooking is a wholly artificial and unnatural process," Macfadden wrote in the *Encyclopedia*, and while he admitted that some foods needed to be cooked—bread, for instance—raw food was generally superior. Macfadden taught that a "natural diet" comprised primarily of raw fruits, vegetables, and nuts was an excellent tonic for exhaustion and sleeplessness. In the months since my first foray into living by the P.C. Diet, both of those plagues had been visited upon me, along with some of the weight I'd lost. For two weeks, I decided, I would eat 99 percent raw.

Few dietary cults are more fanatical than raw foodists. Cooked comestibles are almost always mocked by them as "dead food," and supposedly anything heated to above 116 degrees Fahrenheit loses essential enzymes that may or may not be the wellsprings of long life and contentment. The science behind these claims is sketchy. Surfing the Web one is likely to find raw food rhetoric such as "You wouldn't set fire to your house—why set fire to your food?"

A diet of uncooked foods is basically a diet of exclusion. Anything pasteurized is off-limits, as is coffee, which in its most common form is doubly bad, as it requires both roasting and brewing. Beer and hard liquor (brewed and distilled) are out of the question, as are some wines. Rice, pasta, cheese, almost all cereals and soups—verboten. The list that I took with me to Whole Foods on my first morning of raw fooding consisted of: organic fruits and vegetables; raw, unsalted nuts; unpasteurized honey; green tea (to make sun tea); some whole grains that could be softened by soaking in cold water; and a few varieties of legumes that I would need to sprout to render edible. This is a process by which beans are moistened and left in the sun to start growing into plants. It is about as appetizing as you might imagine.

The first day wasn't bad. I snuck a small cup of coffee in the morning to stave off headaches and ate a bowl of berries with twice-soaked buckwheat, the texture of which can charitably be described as gooey. You know how small children chew with their mouths open, as if their jaws were hinged, when forced to eat something they don't want? I found myself doing that a lot.

Days two and three were a test. On Sunday, I was hosting a barbecue, and the only way I could avoid wolfing down hot dogs and burgers was to self-medicate with two martinis. I consoled myself with the fact that the olives were cured, not cooked. (I, having consumed nothing but fruit all day, was pickled.) Monday I had a business lunch, and luckily my guest was in the mood for sushi. Macfadden loved cod liver oil, so I assumed a sashimi salad was acceptable. White rice had never looked so exotically enticing. I vowed to be pure for the next eleven days.

By day five, my tremendous surge in fiber intake was wreaking havoc on my insides. Things were shifting. I kept thinking of icebergs cracking and falling into the sea. I felt pregnant. But the next day the discomfort subsided. After almost a week without coffee, though, I wasn't sleeping any better. Meals had become chores. My sprouted legumes tasted like grass. By the end of the first week, more than half of my calories were coming in the form of fresh salsa, avocadoes, and dried cranberries.

I couldn't deny that below the neck, my body was definitely responding to the diet. Morning runs felt great, smooth, and easy. On day eight, I noticed that a fruity smell, like strawberries or freshly chopped cilantro, seemed to be following me. Only after a long, sweaty run did I realize that the odor was me, or rather my perspiration, which had changed chemically. My skin changed, too, clearing up as it had with the fast, but this time it took on a youthful suppleness. (I guess you could call it a glow.) I wasn't sleeping any longer than I had previously, but I wasn't crashing in midafternoon, either. I woke up every day at five, ready to go.

By day eleven, the mere scent of sprouts caused warm, pre-

vomit saliva to flood into my mouth. (My own green-apple bou-
quet, of course, never grew tiresome.) Losing weight on a raw
diet is easy, not least because given the choice between my hun-
dredth handful of organic raisins or just skipping a meal, I often
picked the latter. Macfadden was right: Divorced from spices and
familiar flavors, you do lose your cravings surprisingly quickly.
After fourteen days, I'd dropped from 186.8 pounds down to
178.2. I did feel somewhat better, but I doubt I'd have the forti-
tude to go raw again.

EXPERIMENT 3: KEEP ON WALKING

Walking was Macfadden's preferred mode of exercise. He frequently walked to his office in Manhattan from his first home
across the Hudson River, a distance of 25 miles, and for years he
sponsored long group walks from New York City to his Dansville, New York, health hotel, 325 miles away. Every year, inexpe-
rienced hikers would join and inevitably someone would lose an
extraordinary amount of weight, twenty pounds or more in just
two weeks. Walking, he felt, cleaned the blood, promoted deep
breathing, and stimulated the peristaltic mechanism. ("A long
walk is almost always followed by a desire to defecate," he wrote.)

I had almost no experience in walking for exercise, which always seemed to me something old people in laxative commercials
did while wearing matching track suits and visors. (And by Macfadden's reasoning, if they did all that walking, they wouldn't
need a laxative, right?) Macfadden, though, made walking sound
almost magical. "After every long walk the body will be partly
born again, because many new cells will have been built up," he
wrote. "On a walk from New York to Chicago, for instance, one
would arrive with a body quite different from that with which
one started." I didn't have a hole in my schedule that would allow
a ramble to the Great Plains, so I wanted to see what I could accomplish in a month. Macfadden's walking plan is simple: Walk
as far as you can, rest, and then walk some more. I decided to
walk two hours a day while eating as I normally ate.

On day one, I weighed 191.2 pounds. (My ability to regain weight almost instantaneously in between these regimens should give the reader some idea of what my regular diet is like.) Sometimes I walked two hours in the morning before work. Sometimes I walked at lunch, around Central Park. Walking uses different muscles than running, I noticed; any soreness I felt was in my butt rather than my thighs. Another difference between walking and other forms of aerobic exercise is that you can walk hard every day. Any stiffness disappears quickly. Macfadden noted that walking raises the metabolism, and I noticed that I was usually famished around lunchtime. Which was probably why, after eleven days and twenty-two hours of exercise I had lost exactly one pound.

I looked up Macfadden's *Encyclopedia* entry on proper walking form and noticed something I'd missed. A proper pace, he suggested, was at least 3.5 miles per hour, though preferably no faster than 4 mph—a speed more like a slow jog. I upped my pace and immediately covered at least 25 percent more distance on each tour. Halfway through the month, I started to notice changes in my body. My posture had improved. I felt sleeker and more aware of where I was in relation to other objects. If I timed my walks to counteract my hunger pangs as Macfadden recommends, the cravings disappeared after about twenty minutes.

Thirty days later, I'd lost 4.8 pounds and my pants were definitely baggier. I was certain that I'd have lost more if I hadn't gone home each night and eaten like a blacksmith. Which made me wonder—what if I tried to live by all of Macfadden's ten commandments at one time? What would happen then?

EXPERIMENT 4: THE GRAND P.C. FINALE

As I approached the last regimen, I was in probably the saddest shape I'd ever been in. Eighteen months had passed since my first fast. I had been working at home, which is a constant invitation to peek in the refrigerator, and the subconscious knowledge that I would try at least one more Macfadden regimen gave me carte blanche to eat like Robert De Niro preparing to play the fat Jake LaMotta in *Raging Bull*. After a stressful spring and summer, eased by an extra drink or two and maybe a bowl of ice cream in the evening, I'd managed to work my way back up to 193.4 pounds. Now my pants were uncomfortably tight. I'd noticed to my horror that I was becoming one of those men who can cross their arms and rest them on their bellies, like a shelf. My health wasn't great, either. I'd suffered three bad stomach bugs in three months, a sign that something nasty might have snuck into my system. This book was long overdue, and I was waking up in the middle of the night with anxiety pains between my shoulder blades and in my lower back, pains so sharp that I occasionally had to take a muscle relaxant or forego sleep. My knees ached, so I'd stopped exercising.

I knew that I needed to start with a fast. The question was how long. I'd completed two more three-day water fasts since my first one. Like most unpleasant tasks with positive outcomes they became easier with repetition. I loved the rebooting effect; in three days my body felt cleaner and my bad habits seemed to vanish, forgotten and forgiven. When treating patients, Macfadden often recommended fasting until one's natural hunger returned, which could take weeks. True hunger, he said, was felt in the throat, like thirst, not in the stomach. In cases where no acute malady was being treated, he preferred fasts of six to seven days rather than waiting for signs of appetite.

I thought I might be able to handle a week without food. I wondered if seven days off food would clean out the invader that had been bothering my intestines. I'd also been suffering a respiratory problem for five years, the result of kicking up a pile of dust on a building site and deeply inhaling the cloud I created. My lungs felt scorched, and ever since, each time I got a bad cold, it moved down to my chest and lingered for weeks. Macfadden swore that fasting cured most lung ailments, including his own tuberculosis. Most of all, I was intrigued by stories of fasters who felt a euphoric surge of energy on the fourth or fifth day.

To speed up the cleansing process, I planned to walk two hours each day during the fast. Macfadden claimed to have once lost fifteen pounds in a week by fasting and walking ten miles a day. I'd continue the walking for thirty days afterward, and once back on solids, augment my program by doing Macfadden's spine-strengthening routine with light dumbbells every other day.

Every bite of food that then entered my mouth would be Fletcherized. In his *Encyclopedia* volume on food, Macfadden listed his fifteen Basic Principles of Nutrition. Number nine: "Thorough mastication of food is essential to health." He'd borrowed this idea from Horace Fletcher, a rotund raconteur who discovered that the endless chewing of one's food had salubrious effects. He had slimmed from a portly 205 pounds to a trim 163 in four months by exercising his jaw muscles. The gist was this: Every morsel or sip taken into the mouth—even water—needed to be masticated until it disappeared down the back of the throat, without engaging the swallowing mechanism. Macfadden believed that the process slowed down the eater and filled the stomach, which reduced the amount of food consumed. Fletcherism isn't easy, though. For weeks I tried to get the hang of it, practicing on carrots and apples. At my sister's house a grilled short rib required three thousand chews to mash to a pulp. Eventually I gave up, but promised myself that I'd grind every atom of nutritive matter between my molars at least fifty times before daring to swallow.

For two days prior to starting the P.C. plan, I ate nothing but fruit and vegetables. The morning that my fast began, I read a story in a 1926 issue of *Physical Culture* about a man named George Hasler Johnson who attempted Macfadden's hypothetical walk from Chicago to New York—except he tried it while consuming nothing but water. The already trim Johnson covered thirty miles a day through hilly Allegheny terrain, but had to stop in Pittsburgh because the fat that cushioned the bones in his feet had burned off, making each step excruciating. I could only dream of having such problems.

After my three fasts, I'd become accustomed to the rhythms of those first few days. Lunchtime is the hardest, because that's the hour of the day when feeding is most tightly programmed. Sleep patterns go haywire, and a general sluggishness descends. The tongue becomes coated with a yellowish film. In my case, no matter how often I showered, by the third day I generally smelled like an understaffed ward in a Veteran's Administration hospital. This time around, at the end of the third day, I drank a glass of seltzer with a squeeze of lemon—OK by Macfadden's rules—and decided that I was in pretty good form for someone who hadn't eaten in seventy-two hours. I'd felt no chills and hadn't needed more than one nap per day. On the fourth day, new territory for me, I felt strong enough to work and walked two and a half hours. Assuming that all toxins had left my system, I waited for the clouds to part, revealing Nirvana. My nose mysteriously started to drip, then stopped.

On day five, I was certain that physcultopathic rapture awaited just around the corner. My hunger had long since passed. My new office overlooked a soul-food restaurant, however, so the scent of pulled pork and collard greens wafted through my open window all day. I had been sweating a lot—my perspiration now had no scent at all, even when my shirt was drenched—and felt woozy, so I ate a quarter teaspoon of salt. I tried to do the *New York Times* crossword puzzle but gave up after solving one clue in twenty minutes.

That night I awoke at one a.m. with a brutal headache. I almost never get headaches. Two hours later, the pain was only worse. Macfadden, of course, wouldn't have approved of taking a pain reliever, and I wasn't sure what effect two Tylenol would have on my weirded-out system. I flipped through the *Encyclopedia*, to the pages where Macfadden said that if a patient is suffering horribly, give him a little fruit juice. I drank a half-cup of pomegranate juice and felt marginally better. In the morning, I drank another small glass of juice, and my hunger returned almost immediately. My digestive system, dormant for nearly a week, switched on as if it had been plugged in. Faced with slogging through another thirty-six hours without nutriment, I punted. Euphoria wasn't in the cards.

The consolation prize wasn't bad, though. On days seven and eight, I woke feeling wonderful. Ideas that had been hidden behind fog in my brain for months burst to the surface, and I scribbled them down during the long stretches between walks and sips of juice. For three days, I drank small amounts of juice, then upgraded to juice with pulp. By the end of day twelve, I could eat fruit and vegetables and drink milk. When I weighed myself, I was down to 180 pounds. My belt was two notches tighter.

At this point, Macfadden often recommended a switch to an all-milk diet. I skipped this advice for two reasons: I didn't have access to the raw, unpasteurized milk that he absolutely insisted upon for his patients (his reason—that magical enzymes were being cooked out—sounds a lot like the rhetoric of the raw fooders); and I don't really like milk all that much. I vowed to drink a glass with every meal. About two weeks into the P.C. plan, I felt strong enough to start the strength exercises.

The fat melted off my body in phases, as if I were a chubby celebrity being Photoshopped for the cover of a women's magazine. First my thighs became thinner. Then my cheekbones came out of hiding. Finally, my baby beer belly retreated. After twenty days, my butt was like a rock and my waist was shrinking rapidly, as the fat around my love handles released its death grip.

During my third week, I adopted Macfadden's beloved two-meal-a-day plan, which worked just about the way he said it would. I found I could hold off on eating until 10:30 a.m., by which time I would gladly devour just about anything except sprouted chickpeas. (Macfadden was right about milk as a satisfying food, incidentally; if you drink it in the manner he demands, minuscule sips chewed repeatedly, a glass of the stuff can feel like a meal.) By gorging myself on fruit and other low-fat, high-fiber food, I was able to stave off extreme hunger until five—by which time I could definitely feel it in the back of my throat—or even later if I took a late afternoon walk. The time I spent masticating every mouthful of my healthy brunches and early-bird dinners fifty times gave my stomach a chance to fill up. I went for a run every other day and felt no aches the next morning. The strangest difference I noticed was in my flexibility—I could touch my toes, which I hadn't done in years, and then reach my palms to the floor, which I'd never done.

By the thirty-seventh day, I was down to 175.6 pounds (I'd gained some muscle, too) and had lost four inches from my waist. The custom suit I'd long since abandoned again fit like a glove. My energy level was high and—this, to me, was the most interesting part—my digestive bug and lung problems disappeared. Six months later, during which I've again strayed far from Macfadden's ideals, I've seen no sign of either. Which has me thinking: I wonder if there's any place near my house that sells raw milk?

ACKNOWLEDGMENTS

All authors stand on the shoulders of those who come before them, but this story of Bernarr Macfadden would have been impossible to tell without the loads borne by several people whose interest in the Father of Physical Culture long predated my own.

My deepest thanks go out to Esther Ernst, widow of the late scholar Robert Ernst, who, when I called her out of the blue in the fall of 2003, mentioned that yes, she was pretty sure her husband had left behind a few papers from his Macfadden book, and invited me out to her house to borrow them. What I received was three boxes of note cards, scribbled front and back in her late husband's handwriting, alphabetized by subject: a roadmap to sources only a history professor would think of consulting, a list of dead ends that roughly coincided with my preliminary list of places to begin my search, and interviews with Macfadden associates long departed. Mine would have been a much less colorful book without this trove to pick through.

Another unexpected bonus was the collection of interviews and notes from Professor William Taft's unpublished biography of Macfadden, which he researched in the late 1960s and early '70s. Included within were invaluable transcripts of talks with

Macfadden Publications employees now long dead. The Taft papers are kept in the Todd-McLean Physical Culture Collection at the University of Texas at Austin. This extraordinary repository of books, periodicals, and images is overseen by professors Jan and Terry Todd, a husband-and-wife team who have amassed the world's greatest collection of research materials pertaining to physical fitness. The Todds opened the doors of both their library and their beautiful country house to me, and offered powerful scholarly insights that have informed this book. Their associate Kim Beckwith also provided assistance above and beyond the call of duty.

Brewster Macfadden, the last of the purebred physcultopathists, is a private man who until now has never spoken about his father for publication. He and his wife, Peg, welcomed me into their home, shared memories of Bernarr Macfadden that transformed the image I'd created of him in my head, and then took me out for a nice lunch. Bruce, your father was wrong—the sons of great men sometimes turn out to be pretty great in their own way.

Jim Bennett, curator of the bernarrmacfadden.com Web site, provided invaluable links to information and served as a beacon drawing the dwindling number of people left who remember Bernarr Macfadden. One such person, Richard Derwald, sent me an e-mail one day asking if I might be interested in seeing a bunch of old papers that Macfadden's public-relations man had given him in the 1960s. Those papers, limited though they are to the final years of Macfadden's life, are the only known collection of his correspondence, and they offer invaluable insights into his mind as a businessman. Richard and his wife, Maureen, were perhaps the biggest fans this project had, and their enthusiasm was infectious.

While I doubt that anyone on his deathbed has whispered, "If I had it all to do over, I'd have spooled more microfilm," the Microforms Reading Rooms at the New York Public Library and Yale University's Sterling Memorial Library are two of the better places to pass a few sunny spring afternoons scanning hundred-year-old copies of *Physical Culture*. David Smith and Wayne Furman, gatekeepers to the New York Public Library's amazing Wertheim Study, were generous with their access to this precious resource. Melanie James and Joanne McIntyre at the Library of the General Society of Mechanics and Tradesmen provided invaluable research assistance (and really, how many libraries remain, especially in midtown Manhattan, where a librarian will call you at home to ask if you left your watch behind?). Laura L. Carroll was of great assistance in navigating the American Medical Association's enormous Historical Health Fraud and Alternative Medicine Collection. John Boniol of Cumberland University (which absorbed Castle Heights Military Academy in the 1980s) shared newspaper clips from the fascist cadets' visit to Tennessee. Joe Sapia provided information that rounded out my understanding of Physical Culture City. The Chicago Historical Society's librarians were helpful with information about the Healthatorium and double-checked the name of the heavyweight Macfadden had wrestled in St. Louis. The research department at *Time* graciously allowed me to rummage through their yellowing clippings on Macfadden. Cynthia Cathcart, empress of the Condé Nast Library, provided essential leads and sources.

My father, David Adams, and his wife, Mary McEnery, offered limitless food and shelter when I was tracking down facts in Chicago. My generous in-laws Fred and Aura Truslow were kind enough to watch my children and wise enough to buy a house within walking distance of Georgetown University, where I spent many a day sneezing my way through the dusty scrapbooks and papers of the Fulton Oursler Collection. In both Washington, D.C., and New York, the inexhaustible Natividad Huamani freed up many hours so that I could work. *Mil gracias,* Nati.

One funny thing about trying to write a biography on weekends and during vacations is that it takes a bit longer than you might expect. I've been lucky enough to have been blessed with three great editors along the way. Mark Bryant, one of the least overtly excitable humans on the planet, got the ball rolling over lunch one day by raising an eyebrow when I told him Macfadden's life story. His advice in shaping the original proposal for this book, and his goading to hammer out a workable outline after I turned in one that, as Mark might say, "needed to be run through the typewriter again," was invaluable. John Williams, my second editor, read the first half and said exactly what I needed to hear: "This is great—keep going." Allison Lorentzen went over the manuscript with the eye of a lepidopterist; like a true kinisitherapist, she pointed out the flabby spots and requested that I pack on a bit more factual muscle. If the final product is in half as good shape as Macfadden was, I have her to thank. Gary Stimeling's copy edits saved me from committing some potentially mortifying errors.

Numerous friends agreed to read the manuscript in rough form. David McAninch provided his usual insightful analysis and helped me recover from my fortnight of raw foodery with a ten-thousand-calorie lunch. Maura Fritz, arguably the finest line editor of her generation, took a very large chunk of time out from a very busy year to provide her customary level of polish. Gillian Fassel, whose surgical skills with a blue pencil could have converted Macfadden to the side of vivisection—I can think of no one I'd rather have scribble "zzzzz" in the margin next to a paragraph it took a week to write—palpated this body of work until it gave up its secrets. Adam Brightman, Steve Byers, and Charlie Vanek read the manuscript in draft form and provided new perspectives. Professor Jacqueline Reich was kind enough to fact-check my Mussolini tale and to not gloat at the mistakes she found.

During the writing of this book, I was blessed to work for bosses with a deep respect for the written word, who didn't complain when I came stumbling back to the office from the occasional three-hour lunch at the library, shamelessly reeking of toner: Susan Casey, John Rasmus, and Bill Shapiro. My agent, the unflappable Daniel Greenberg, employed his uncommonly well-tuned ear for prose to harmonious ends and demonstrated his empathy by going on a fast of his own. It's hard enough to find an agent who'll give his time for an author; how many would give up their food?

Those who listened to me complain about this book over the years with a sympathetic ear include: John Hodgman, who offered insights into the world of publishing and is lucky to be married to Katherine Fletcher, believed to be an actual blood relation of the Great Masticator; nonobservant Cosmotarian Brett Martin; Devin "Isn't that book done yet?" Friedman and Danielle Pergament; codebreaker Edward McPherson; Sheldon Moyer; Lauren Thomas; Kerry McNicholas; Adam Sachs; Tom Mallon; Dan Ferrara; Veronica Francis and Pat Manocchia, who taught me to like exercise at the age of thirty, even if he couldn't teach me to catch a football properly. My brother, Jason Adams, has been an endless source of support, emotional, editorial, and viticultural. Sarah Adams, my youngest sister, kindly snapped my author photograph, gloveless, on a fifteen-degree day in Brooklyn.

The march of time limited the number of people whom I was able to interview for this book, but some of the healthier folks still around to share their impressions of Macfadden included Jack La-Lanne, Jack Macfadden, Dan Wiegers, and Fulton Oursler Jr. and his wife, Noel, who were kind enough to lend me their personal copy of *Chats with the Macfadden Family*. Ernest Rubenstein of Paul, Weiss described the Macfadden Foundation's demise. I owe a deep debt of gratitude to those who welcomed me to Dansville, New York, and granted me access to the former Physical Culture Hotel: Al Jamison, Don Sylor, and Mayor William Dixon. Joseph Elliott and his wife, Victoria, who worked at the hotel in the 1940s, granted me several hours of their time before Joe had to leave to teach a ballroom dancing lesson. Local archivist Jane Schryver shared hundreds of photos of the hotel at its best and worst.

Dozens of people e-mailed me to reminisce about their time attending Macfadden schools or to boast about how their parents had made them follow Macfadden-approved remedies. Your stories have helped inform this narrative. Many more e-mailed trying to sell me Macfadden memorabilia; I hope eBay was kind to you.

My greatest debt in all things is to my wife, Aurita, whose love and support have sustained me throughout this project. Honey, I promise to never sprout legumes in the kitchen again.

And for their patience during a lot of Saturdays in which Daddy was at the office writing, a special thanks to my boys, Alex, Lucas, and Magnus.

NOTES

At the end of his biography of the Jesuit martyr Edmund Campion, Evelyn Waugh wrote of the cleric's life: "There is a great need for a complete scholar's work on the subject. This is not it. All I have sought to do is to select the incidents which strike a novelist as important and put them in a narrative which I hope may prove readable." Let's just say that when I read those words I had the same experience Macfadden did reading William Blaikie's *How to Get Strong and How to Stay So*. Every sentence in this book has a factual basis. If a specific detail isn't documented in the notes below and you want to know where it came from—or if you spot an error—please e-mail me at bernarrmacfadden@gmail.com.

Several key texts were important sources for virtually every chapter in this book:

Weakness Is a Crime: The Life of Bernarr Macfadden, by Robert Ernst. Probably as close to a definitive scholarly biography as we'll see, unless a cache of Macfadden papers turns up in a trunk somewhere.

"My Life Story" and "My Fifty Years of Physical Culture," by Bernarr Macfadden. These multipart serials, which cover much parallel material, ran in *Physical Culture* in 1914–15 and 1933–34. In them, Macfadden told the story of his boyhood and discovery of physcultopathic methods. The later serial, written when Macfadden was polishing his résumé for a political career, is less entertaining and informative than the first, but the essential facts are the same.

The True Story of Bernarr Macfadden, by Fulton Oursler. The slickest and most informative of the three Macfadden biographies published in 1929, written by his friend and editorial director.

Bernarr Macfadden: The Muscular Prophet, by Clifford Waugh. This dissertation, written in 1973 by a PhD candidate at the State University of New York–Buffalo, is perhaps the best review of Macfadden's career. Waugh spoke with many of Macfadden's closest advisers and did an extraordinary amount of original research.

Dumbbells and Carrot Strips, by Mary Macfadden with Emile Gauvreau. If Kitty Kelley had been unhappily married to America's top health guru, this is the memoir she would have written. It's loaded with details about the Macfadden family home life.

Physical Culture magazine, 1899–1941. Macfadden's baby. It truly does have to be seen to be believed.

I am also deeply grateful for the work of the long-departed Macfadden-era reporting staffs of the *New York Times*, the *New York Evening Graphic*, *Time*, *Newsweek*, *Editor and Publisher*, the *New York Herald Tribune*, the *Bergen County Record*, the *Chicago Tribune*, and the *Los Angeles Times*. Without their work, writing this book would have been far less entertaining.

The roundup of general sources that follows is by no means exhaustive. I've left out most quotes that are attributed in the text, and almost all newspaper and magazine stories, a list of which would fill an entire book every bit as fascinating as *The Athlete's Conquest*. It's safe to assume that *Physical Culture*, Macfadden's Editor's Viewpoint columns in particular, pervades every chapter.

PROLOGUE

The details of Macfadden's trip to the CBS studio came from a memo in a collection of Macfadden's private papers from the years 1941–55, owned by Richard Derwald, and from an interview with Macfadden's grandson Dan Wiegers, who watched the broadcast in 1951. Other details were plucked from *What's My Line?* producer Gil Fates's memoir *What's My Line? TV's Most Famous Panel Show*. Bennett Cerf's diary, buried among his papers at Columbia University, revealed that he was the panelist who'd identified Macfadden. Sadly, the kinescope of Macfadden's appearance was erased.

CHAPTER 1

Robert Lewis Taylor's three-part profile of Macfadden appeared in the October 14, 21, and 28, 1950, issues of the *New Yorker*. (Fun fact: The collection of *New Yorker* editor Harold Ross's correspondence, *Letters from the Editor*, edited by Thomas Kunkel, contains a note to James Thurber inquiring about a story on the "Macfadden for President" campaign; alas, no such article exists.) The history of Missouri was cobbled together from several sources, including Lucien Carr's *Missouri, a Bone of Contention*, Paul C. Nagel's *Missouri: A History*, and Professor William H. Taft's essay "Bernarr Macfadden" in the

Missouri Historical Review. Details of Macfadden's childhood come from all the major biographies and memoirs, which cover much of the same ground, but especially Macfadden's 1914 *Physical Culture* series and Taylor's *New Yorker* trilogy. (The irresistible quote about Macfadden's being traded for a "scattering of mixed produce" is Taylor.) Negative impressions of Macomb, Illinois, are based on the author's personal experience.

CHAPTER 2

The history of St. Louis draws on Ernest Kirschten's *Catfish and Crystal*. The location of the St. Louis Gymnasium is based on a Library of Congress map to be found at http://memory.loc.gov/cgi-bin/map_item.pl. I plucked the excellent Ben Franklin quote from an essay by Jan Todd in her journal, *Iron Game History*, "Strength Is Health: George Barker Windship and the First American Weight Training Boom." (Articles on Windship, Lewis, Blaikie, and other fitness pioneers can also be found in *IGH*.) Much of the history of fitness in the nineteenth century is told in two excellent sources: James Whorton's *Crusaders for Fitness* (in which Whorton compares Dudley Allen Sargent to Isaac Newton) and Harvey Green's *Fit for America: Health, Fitness, Sport and American Society*. Macfadden's admiration for Jane Austen was mentioned by his daughter Beverly in a letter to Professor William H. Taft. Almost all of the detail in this chapter about Macfadden's teenage years is taken from his 1914 memoirs in *Physical Culture*.

CHAPTER 3

Details from the 1893 Columbian Exposition come from numerous sources, including Erik Larson's *The Devil in the White City* and Austin Hoyt's PBS documentary *Chicago: City of the Century*. Details of Sandow's life and career are from John Kasson's *Houdini, Tarzan and the Perfect Man: The White Male Body and the Challenge of Modernity in America* and David Chapman's *Sandow the Magnificent: Eugen Sandow and the Beginnings of Bodybuilding*. The thumb's-up review of Eugen Sandow's performance at the Trocadero appeared in the August 11, 1893, *Chicago Tribune*. (Further details on the history of bodybuilding can be found at the comprehensive www.sandowplus.co.uk Web site.) Descriptions of Macfadden's early career are taken from his 1914 and 1933 serials. Macfadden's first wife, Bertha Fontaine, has been identified as "Tillie Fontaine" by Robert Ernst and others; in a tiny 1946 story that I found buried in the *Los Angeles Times*, an unwell Bertha Fontaine Tilley pleaded for help and admitted that she was Macfadden's first wife.

CHAPTER 4

The figure of eighty-five health magazines in the United States came from Gerald Carson's delightful history *Cornflake Crusade*. The quote about feminine beauty and muscles is from Jan Todd's essay "Bernarr Macfadden: Reformer of Feminine Form," originally published in the *Journal of Sport History*. The Nick Carter description, taken from an early Carter story, is in Robert Sampson's *Yesterday's Faces: A Study of Series Characters in the Early Pulp Magazines*. The estimate of *Physical Culture* profits is Macfadden's own, from 1914. Background information on America's diet at the turn of the century can be found in Harvey Levenstein's excellent *Revolution at the Table: The Transformation of the American Diet*. (His equally fine *Paradox of Plenty* was also useful.) Footage from the film that Thomas Edison's production company shot of Macfadden's 1903–4 exhibition can be found at the Library of Congress Web site. Details about the second exhibition are taken primarily from Macfadden's 1914 memoir and the *New York Times*. The best biography of Anthony Comstock is probably Heywood Broun and Margaret Leech's 1927 book, *Anthony Comstock: Roundsman of the Lord*. The shaping of this chapter, especially, was aided by the lucid analysis found in Clifford Waugh's dissertation *Bernarr Macfadden: The Muscular Prophet*.

CHAPTER 5

The section on the Progressive Era and the muckrakers who helped drive it was informed by Richard Hofstadter's book *The Age of Reform* and Theodore Peterson's *Magazines in the Twentieth Century*. Clifford Waugh conducted essential interviews with Physical Culture City alumni. Allan Brandt's *No Magic Bullet: A Social History of Venereal Disease in the United States since 1880* provided much of the background on that formerly taboo topic. The quote from Mary Macfadden is in *Dumbbells and Carrot Strips*.

CHAPTER 6

The complete *Encyclopedia of Physical Culture*, originally called *Macfadden's Encyclopedia of Physical Culture* and later renamed *The Encyclopedia of Health*, appeared in multiple editions starting in 1912. The definitive history of Battle Creek's heyday as a health resort is from Gerald Carson's *Cornflake Crusade*. James Whorton's *Crusaders for Fitness* and the essay collection *Other Healers* (edited by Norman Gevitz) provided much historical background on early alternative health practitioners. (Many, if not most, of those practitioners also told their own stories in *Physical Culture*.) Upton Sinclair described his time with Macfadden at length in both *American Outpost* and his *Autobiography*. Jean Toomer's impressions of physcultopathy are collected in *The Wayward and the Seeking*.

CHAPTER 7

The quote from Charles Benedict Davenport's book is cited in Ruth Clifford Engs's *Clean Living Movements: American Cycles of Health Reform*. Almost all the detail of the Perfect Woman contest and the subsequent marriage of the

Macfaddens draws on *Dumbbells* (which is why Mary's wit may seem unusually sophisticated for a nineteen-year-old) and Macfadden's 1914 serial.

CHAPTER 8

Vitality Supreme is almost certainly Macfadden's finest book; not only does it condense physcultopathy to its essence, it also offers insight into the years when "efficiency" became an American obsession. John D'Emilio and Estelle B. Freedman's history of sex in America, *Intimate Matters*, helped me to unravel the early years of the sexual revolution.

CHAPTER 9

The quote about all love stories being about sex is from Stanley Walker's collection of profiles *Mrs. Astor's Horse*. Frederick Lewis Allen's book *Only Yesterday* is an invaluable resource for anyone trying to fathom the giddy mood of America in the 1920s. Mary Macfadden gives her version of *True Story*'s creation in *Dumbbells*; other variations can be found in Harold Hersey's *Pulpwood Editor* and Fulton Oursler's *The True Story of Bernarr Macfadden*. Information about the Big Six is from Mary Ellen Zuckerman's *A History of Popular Women's Magazines in the United States 1792–1995* and Peterson's magazine history. Dr. Reeder's assistance in conceiving a son is described—from very different points of view—in *Dumbbells* and Bernarr Macfadden's many articles in *Physical Culture* crowing about the feat. Much of the description of the Fortieth Street offices is from Fulton Oursler's memoir, *Behold This Dreamer*. John Fair's *Muscletown USA*, a biography of *Strength and Health* founder Bob Hoffman, doubles as an excellent history of weightlifting and bodybuilding in America. The tale of Charles Atlas's early involvement with Macfadden is drawn largely from articles in *Physical Culture*, but Charles Gaines and George Butler's *Yours in Perfect Manhood, Charles Atlas* helped place the strongman in context, as did Robert Lewis Taylor's 1942 *New Yorker* profile and Frederick Tilney's autobiography *Young at 73—and Beyond!* Oursler's *Behold This Dreamer* is the primary source for background on his joining Macfadden's company; the source materials for *that* book, kept in dozens of boxes at Georgetown University Library, provided many small details. It was Alva Johnston who called confessions magazine publishing the "I'm Ruined! I'm Ruined" industry, in the *Saturday Evening Post*. The description of Macfadden's "gang war" hat came from a letter written by his daughter Beverly, in William H. Taft's papers at the University of Texas. The most thorough telling of Oursler's early years at Macfadden Publications can be found in *Behold This Dreamer*. Details of Byron Macfadden's death can be found in *Dumbbells*, as well as in his father's postmortem editorial in *Physical Culture* and Oursler's recollections in *Behold*. Circulation figures for Macfadden Magazines are taken from advertisements Macfadden placed touting the stock in his magazines and from the Audit Bureau of Circulations.

CHAPTER 10

Gauvreau's quote is from his memoir, *My Last Million Readers*, as are some of the details of his start at the *Graphic*. Two books provided background on the golden age of tabloid journalism: Simon Michael Bessie's *Jazz Journalism* and Silas Bent's *Ballyhoo*. Many versions of the *Graphic*'s birth have been told. Two books written by former *Graphic* editors were especially useful: contest editor Lester Cohen's *The* New York Graphic: *The World's Zaniest Newspaper* and Frank Mallen's *Sauce for the Gander*. Descriptions of Gauvreau's tenure at the *Hartford Courant* were taken from John Bard McNulty's *Older than the Nation: The Life and Times of the* Hartford Courant. The first several weeks of the *Graphic* are available on microfilm at the New York Public Library, but the library's holdings are patchy for most of the paper's eight-year run; many of the headlines mentioned here now exist only in Cohen's and Mallen's books. No other collection of *Graphic*s is known to exist, though a few hard copies of the paper are kept amid the Oursler collections at Georgetown. (A St. Patrick's Day edition was printed on green paper rather than the usual pink.) George Rosen's classic *A History of Public Health* helped clarify key points about medicine in the first quarter of the twentieth century; James Whorton's *Nature Cures: The History of Alternative Medicine in America* was also helpful. Most of the detail of the war between Macfadden and the American Medical Association is based on letters and clippings from the AMA's Historical Health Frauds file. In *Weakness Is a Crime*, Robert Ernst placed more emphasis on Macfadden's possible role in his daughter Byrne's death. I chose not to, for two reasons: His source was former *Dance Lovers* editor Vera Caspary's book *The Secrets of Grown-Ups*, published more than fifty years after the incident by a person who strongly disliked Macfadden; and Mary barely mentions the death in *Dumbbells*, a book that was largely plotted to make her husband look malicious.

CHAPTER 11

Details of the *Graphic* newsroom can be found in all the books listed above; these are buttressed by biographies of two alumni, Neal Gabler's *Winchell: Gossip, Power and the Culture of Celebrity* and James Maguire's *Impresario: The Life and Times of Ed Sullivan*. The unique No Smoking sign appeared in an August 31, 1929, story about the paper in *Editor and Publisher*. Allen Churchill's *The Year the World Went Mad* was helpful in digesting the craziness of the year 1926. Many of the descriptions of reactions to Valentino's death come from Mallen's *Sauce for the Gander*, but the *New York Times* and Emily W. Leider's *Dark Lover: The Life and Death of Rudolph Valentino* were also consulted. Gauvreau's memo is quoted at greater length in Mallen's book. A. Scott Berg's *Lindbergh* was useful in navigating the news maelstrom created by Charles Lindbergh's historic flight.

CHAPTER 12

Though no book has been written about the history of *True Story*, to my knowledge, Jacqueline Hatton's PhD dissertation *True Stories: Working-Class Mythology, American Confessional Culture, and "True Story" Magazine 1919–1929* gives a fine overview of that magazine's explosive impact on American popular culture. The description of the Macfaddens at the Nyack pool is from the autobiography of another forgotten character of the twentieth century, Cornelius Vanderbilt's ridiculously entertaining *Man of the World*. Edward Bernays devotes a few pages to Macfadden in his doorstop *Biography of an Idea*. (However, Macfadden's pride in his daughter Byrnece's physical form, and details from his publicity tour and 1928 political flight itinerary, are from Clement Wood's authorized biography, *Bernarr Macfadden: A Study in Success*.) The meeting with Olvany is described in Gauvreau's *My Last Million Readers*. The early description of the Dansville resort is from Grace Perkins Oursler's *Chats with the Macfadden Family*. Lee Ellmaker's arrival at the *Graphic* was described in Cohen's *Graphic* history. The investment prospectus was 1929's *Statistical and Financial Analysis of the Macfadden Publications*. Details of Winchell's firing are from Gauvreau and from Lyle Stuart's *The Secret Life of Walter Winchell*, as quoted in Gabler's biography. The story of Macfadden's visit with Mussolini is told in *Behold, Dumbbells,* and numerous *Physical Culture* articles; most of the information, and that about the Italian cadets' trip to America, is from a book Macfadden published titled *Italian Physical Culture Demonstration* by Thomas B. Morgan.

CHAPTER 13

Details about Macfadden's purchase of *Liberty* and Oursler's home offices come from Fulton Oursler's *Behold* and his son Will Oursler's *Family Story*. The letter from Macfadden to Thomas Dewey is in the former governor's papers at the University of Rochester. Cornelius Vanderbilt's recollections are from *Man of the World*. Several biographies of the Roosevelts were helpful in preparing the section on their dealings with Macfadden and *Liberty*, including Joseph Lash's *Eleanor and Franklin*, and Conrad Black's *Franklin Delano Roosevelt*. (Oursler's *Behold* also contains a long section on his work with the Roosevelts.) Details of the Macfadden Foundation's finances were published in many major newspapers the day after the announcement. The description of the Englewood home is based on those in *Chats with the Macfadden Family*. Byrnece's reminiscences of her father's diets were related in a letter to William H. Taft. Details of spying in Englewood come from letters written by Macfadden

Publications employee Marjorie Greenbie, kept among the papers of her husband Sydney at the University of Oregon—which I never would have known about if Robert Ernst hadn't mentioned them in his notes. Byrnece Muckerman shared her father's prewedding advice with Clifford Waugh. Many details of the Macfaddens' battles and separation are taken from their divorce papers, which were filed in Florida in 1945. The $11 million loss is cited in Mallen's *Sauce for the Gander.*

CHAPTER 14

Much of the background on the 1930s and the Great Depression is from Frederick Lewis Allen's *Since Yesterday* and contemporaneous news accounts. The description of Sandalwood and Oursler's satellite office come from *Behold* and an interview with Fulton Oursler Jr. All stock prices are from Moody's annual reports. Orr Elder Jr.'s quote is from Clifford Waugh's *Muscular Prophet.* The rumor about the photo studio chief is from an interview Robert Ernst conducted with Macfadden Publications executive George Davis. The passenger who flew with Macfadden was S. O. Shapiro, also interviewed by Ernst. (A brief video clip of Macfadden flying the Hughes plane can be seen on the Web site for Pathé Films, www.britishpathe.com.) The *Physical Culture* cartoon by George Price appeared in the July 29, 1933, issue of the *New Yorker.* Details of the Portuguese children's colony come from the book *A Dreamer in Portugal,* written by Thomas Dixon and paid for by Macfadden. Jack LaLanne's introduction to Paul Bragg is from an interview with LaLanne by the author. The "Rear Admiral" nickname was told to the author by Joe Elliott. The sample menu and hotel brochure in which Lowell Thomas's quote appeared are in the Derwald collection. The quote about the Deauville's bars was written in a letter from former Macfadden Publications employee Barry Good to Clifford Waugh.

CHAPTER 15

Information about Mary Macfadden's possible attempt to frame Oursler for the Lindbergh baby kidnapping is frustratingly scarce. Oursler deals with it at length in *Behold* (whence most of the details here are taken) and Photostats of the letters can be found in the Oursler collection at Georgetown; *Time* wrote about it; Mary Macfadden, interestingly, did not mention it in *Dumbbells.* Scott Berg's *Lindbergh* provides a thorough overview of the murder and its aftermath. Charlton Tebeau's *History of Florida* helped explain politics in the Sunshine State. Almost all of the detail about Macfadden's senate campaign comes from a privately published book, *Confessions of an Amateur Politician,* which the Macfadden Foundation printed in 1948. The photograph of Macfadden jumping over a deck chair appeared in the *Los Angeles Times.* Joe Weider's recollection of lunch with Macfadden appears in his autobiography,

Brothers of Iron, written with his brother Ben Weider and Mike Steere. The story of bakers putting vitamins back into bread is told in Harvey Levenstein's *Paradox of Plenty* and far too many *Physical Culture* stories to count. Details of Macfadden's alleged misdeeds are taken from the lawsuit *Leon S. Brach* vs. *Macfadden Publications and Bernarr Macfadden*. My description of the coup drew on interviews conducted by Clifford Waugh for his dissertation. Joseph Schultz's quote was recounted to Robert Ernst by Dell Publishing founder George Delacorte. The quote about Macfadden's failure to acknowledge that he ran a public company was from S. O. Shapiro, also as told to Robert Ernst.

CHAPTER 16

The letter to Brewster Macfadden is in Richard Derwald's collection. The letters from J. Edgar Hoover can be found in Macfadden's and Oursler's FBI files. Fulton Oursler Jr. was kind enough to explain what the American Editors Syndicate had been. Almost all of the information on Cosmotarianism is based on materials given to Richard Derwald by Ed Bodin, Macfadden's former public-relations man. The best description of the Carnegie Hall speech appeared in *Newsweek*. Information about the children's trust funds was taken from legal documents among the Derwald papers. His note to Mary is also in that collection. The section on Johnnie Lee Macfadden draws on interviews conducted by Robert Ernst; Johnnie Lee also left papers to the Todd-McLean Collection at the University of Texas. S. O. Shapiro told Ernst the story of hearing Winchell announce Macfadden's (false) impending fatherhood on the radio. Numerous letters and telegrams documenting Macfadden's failed attempt to prove Johnnie Lee was a bigamist can be found in the Derwald papers. The Bernarr Macfadden Foundation's end is something of a mystery. The foundation's president, Joseph Wiegers (husband of the former Helen Macfadden) wrote to William Taft in 1970 that the nonprofit had $6 million in assets. Ernest Rubenstein of the New York law firm Paul, Weiss told me the firm was retained in the 1960s to settle outstanding lawsuits, and that little money was left after settlements. According to the Lexis legal database, the Bernarr Macfadden Foundation was dissolved due to bankruptcy in 1978.

EPILOGUE

Among the many books on general health and fitness history that were especially useful in writing this section were: Scott Mowbray's *The Food Fight: Truth, Myth, and the Food-Health Connection*, James Whorton's *Nature Cures*, and Carlton Jackson's *J. I. Rodale: Apostle of Nonconformity*. My special thanks to the very nice stranger who showed me how to find Macfadden's grave at Woodlawn Cemetery, without whose help I might still be wandering around the Bronx, notebook in hand.

SELECTED BIBLIOGRAPHY

Addison, Heather. *Hollywood and the Rise of Physical Culture*. New York and London: Routledge, 2003.

Allen, Frederick Lewis. *Only Yesterday: An Informal History of the 1920s*. New York: Harper & Brothers, 1931.

Allen, Frederick Lewis. *Since Yesterday: The 1930s in America*. New York: Harper & Brothers, 1940.

Anonymous. *Statistical and Financial Analysis of Macfadden Publications*. New York: Macfadden Publications, 1929.

Arthur, Anthony. *Radical Innocent: Upton Sinclair*. New York: Random House, 2006.

Barnouw, Eric. *A Tower in Babel: A History of Broadcasting in the United States to 1933*. New York: Oxford University Press, 1966.

Bent, Silas. *Ballyhoo*. New York: Horace Liveright, 1927.

Berg, A. Scott. *Lindbergh*. G. P. Putnam's Sons, 1998.

Bernays, Edward L. *Biography of an Idea: Memoirs of Public Relations Counsel Edward L. Bernays*. New York: Simon & Schuster, 1965.

Bessie, Simon Michael. *Jazz Journalism: The Story of the Tabloid Newspapers*. New York: E. P. Dutton, 1938.

Black, Conrad. *Franklin Delano Roosevelt*. New York: PublicAffairs, 2003.

Blaikie, William: *How to Get Strong and How to Stay So*. New York: Harper & Brothers, 1879.

Brandt, Allan. *No Magic Bullet: A Social History of Venereal Disease in the United States Since 1880*. New York: Oxford University Press, 1985.

Broun, Heywood, and Margaret Leetch. *Anthony Comstock, Roundsman of the Lord*. New York: Albert & Charles Boni, 1927.

Carr, Lucien. *Missouri, Bone of Contention.* Boston: Houghton-Mifflin, 1888.

Carson, Gerald. *Cornflake Crusade.* New York: Rinehart, 1957.

Caspary, Vera. *The Secrets of Grown-ups.* New York: McGraw-Hill, 1979.

Chapman, David L. *Sandow the Magnificent: Eugen Sandow and the Beginnings of Bodybuilding.* Urbana and Chicago: University of Illinois Press, 1994.

Churchill, Allen. *The* Liberty *Years: 1924–1950.* Englewood Cliffs, NJ: Prentice-Hall, 1968.

Churchill, Allen. *The Year the World Went Mad.* New York: Thomas Y. Crowell, 1960.

Cohen, Lester. *The* New York Graphic: *The World's Zaniest Newspaper.* Philadelphia: Chilton Books, 1964.

Davis, Adelle. *Let's Eat Right to Keep Fit.* New York: Harcourt, Brace, 1954.

D'Emilio, John, and Estelle B. Freedman. *Intimate Matters: A History of Sexuality in America.* New York: Harper & Row, 1988.

Dixon, Thomas. *A Dreamer in Portugal.* New York: Covici, Friede, 1934.

Engs, Ruth Clifford. *Clean Living Movements: American Cycles of Health Reform.* Westport, CT: Praeger, 2000.

Engs, Ruth Clifford. *The Progressive Era's Health Reform Movement: A Historical Dictionary.* Westport, CT: Praeger, 2003.

Ernst, Robert. *Weakness Is a Crime: The Life of Bernarr Macfadden.* Syracuse, NY: Syracuse University Press, 1991.

Fabian, Ann. *The Unvarnished Truth: Personal Narratives in Nineteenth-Century America.* Berkeley: University of California Press, 2000.

Fair, John. *Muscletown, USA: Bob Hoffman and the Manly Culture of York Barbell.* University Park: Pennsylvania State University Press, 1999.

Fates, Gilbert. *What's My Line? The Inside History of TV's Most Famous Panel Show.* Englewood Cliffs, NJ: Prentice-Hall, 1978.

Fishbein, Morris. "Exploiting the Health Interest," parts 1 and 2. *Hygeia,* Nov. and Dec. 1924.

Fishbein, Morris. *Medical Follies: An Analysis of the Foibles of Some Healing Cults.* New York: Boni & Liveright, 1925.

Fletcher, Horace. *The A.B.-Z of Our Own Nutrition.* New York: Frederick A. Stokes, 1903.

Gabler, Neal. *Winchell: Gossip, Power and the Culture of Celebrity.* New York: Alfred A. Knopf, 1994.

Gaines, Charles, and George Butler. *Yours in Perfect Manhood, Charles Atlas.* New York: Simon & Schuster, 1982.

Gauvreau, Emile. *Hot News.* New York: Macaulay Company, 1931.

Gauvreau, Emile. *My Last Million Readers.* New York: E. P. Dutton, 1941.

Gevitz, Norman, ed. *Other Healers: Unorthodox Medicine in America.* Baltimore: Johns Hopkins University Press, 1988.

Green, Harvey. *Fit for America: Health, Fitness, Sport and American Society*. New York: Pantheon, 1986.

Hale, Annie Riley. *These Cults: An Analysis of the Foibles of Dr. Morris Fishbein's "Medical Follies" and an Indictment of the Medical Practice in General, with a Non-Partisan Presentation of the Case for the Drugless Schools of Healing, Comprising Essays on Homeopathy, Osteopathy, Chiropractic, the Abrams Method, Vivisection, Physical Culture, Christian Science, Medical Publicity, the Cost of Hospitalization and State Medicine*. New York: National Health Foundation, 1926.

Hatton, Jacqueline. *True Stories: Working-Class Mythology, American Confessional Culture, and* True Story *Magazine, 1919–1929*. PhD dissertation, Cornell University, 1997.

Hauser, Gayelord. *Look Younger, Live Longer*. New York: Farrar, Straus, 1950.

Hersey, Harold. *Pulpwood Editor*. New York, Frederick A. Stokes, 1937.

Hofstadter, Richard. *The Age of Reform: From Bryan to F.D.R.* New York: Alfred A. Knopf, 1955.

Hoyt, Austin. *Chicago: City of the Century*. Produced, written, and directed by Hoyt for PBS's *American Experience*, 2004.

Hunt, William R. *Body Love: The Amazing Career of Bernarr Macfadden*. Bowling Green, OH: Bowling Green State University Popular Press, 1989.

Jackson, Carlton: *J. I. Rodale: Apostle of Nonconformity*. New York: Pyramid Books, 1974.

Johnston, Alva. "The Great Macfadden," parts 1 and 2. *Saturday Evening Post*, June 21 and 28, 1941.

Kemp, Harry. *Tramping on Life: An Autobiographical Narrative*. Garden City, NY: Garden City Publishing, 1922.

Kasson, John. *Houdini, Tarzan and the Perfect Man: The White Male Body and the Challenge of Modernity in America*. New York: Hill & Wang, 2001.

Kirschten, Ernest. *Catfish and Crystal*. Garden City, NY: Doubleday, 1965.

Kolata, Gina. *Ultimate Fitness: The Quest for Truth about Exercise and Health*. New York: Farrar, Straus & Giroux, 2003.

Larson, Erik. *The Devil in the White City: Murder, Magic and Madness at the Fair That Changed America*. New York: Crown, 2003.

Lash, Joseph P. *Eleanor and Franklin: The Story of Their Relationship, Based on Eleanor Roosevelt's Private Papers*. New York: W. W. Norton, 1971.

Leider, Emily W. *Dark Lover: The Life and Death of Rudolph Valentino*. New York: Farrar, Straus & Giroux, 2003.

Levenstein, Harvey A. *Paradox of Plenty: A Social History of Eating in Modern America*. Berkeley: University of California Press, 2003.

Levenstein, Harvey A. *Revolution at the Table: The Transformation of the American Diet*. New York: Oxford University Press, 1988.

McCollum, Elmer V. *A History of Nutrition*. Boston: Houghton-Mifflin, 1957.

Macfadden, Bernarr. *The Athlete's Conquest: The Romance of an Athlete*, rev. ed. New York: Physical Culture Publishing Co., 1901.

Macfadden, Bernarr. *Confessions of an Amateur Politician*. New York: Macfadden Foundation, 1948.

Macfadden, Bernarr. *Fasting for Health: A Complete Guide on How, When and Why to Use the Fasting Cure*. New York: Macfadden Book Co., 1935.

Macfadden, Bernarr, and Felix Oswald. *Fasting, Hydropathy, and Exercise*. New York: Physical Culture Publishing, 1900.

Macfadden, Bernarr. *Macfadden's Encyclopedia of Physical Culture: A Work of Reference Providing Complete Instructions for the Cure of All Diseases through Physcultopathy, with General Information on Natural Methods of Health-Building and a Description of the Anatomy and Physiology of the Human Body*. New York: Macfadden Publications, 1912.

Macfadden, Bernarr. *Macfadden's New Hair Culture*. New York: Physical Culture Publishing Co., 1901.

Macfadden, Bernarr. *Macfadden's Physical Training: An Illustrated System of Exercises for the Development of Health, Strength and Beauty*. New York: Macfadden Co., 1900.

Macfadden, Bernarr. *Manhood and Marriage*. New York: Physical Culture Publishing Co., 1916.

Macfadden, Bernarr. *Man's Sex Life*. New York: Macfadden Book Co., 1936.

Macfadden, Bernarr. *Marriage a Lifelong Honeymoon: Life's Greatest Pleasures Secured by Observing the Highest Human Instincts*. New York: Physical Culture Publishing Co., 1903.

Macfadden, Bernarr. *The Miracle of Milk: How to Use the Milk Diet Scientifically at Home*. New York: Macfadden Publications, 1923.

Macfadden, Bernarr. *Muscular Power and Beauty: Containing Detailed Instructions for the Development of the External Muscular System to Its Utmost Degree of Perfection*. New York: Physical Culture Publishing Co., 1906.

Macfadden, Bernarr. "My Fifty Years of Physical Culture." *Physical Culture*, various issues, 1933–1935.

Macfadden, Bernarr. "My Life Story." *Physical Culture*, various issues, 1914–1915.

Macfadden, Bernarr, with Milo Hastings. *The Physical Culture Cook Book*. New York: Macfadden Publications, 1929.

Macfadden, Bernarr. *Strength from Eating: What and How to Eat and Drink to Develop the Highest Degree of Health and Strength*. New York: Physical Culture Publishing Co., 1901.

Macfadden, Bernarr, and John R. Coryell. *A Strenuous Lover: A Romance of Love's Vast Power.* New York: Physical Culture Publishing Co., 1904.

Macfadden, Bernarr. *Strong Eyes.* New York: Physical Culture Publishing Co., 1901.

Macfadden, Bernarr. *The Virile Powers of Superb Manhood: How Developed, How Lost, How Regained.* New York: Physical Culture Publishing Co., 1900.

Macfadden, Bernarr. *Vitality Supreme.* New York: Macfadden Publishing Co., 1915.

Macfadden, Bernarr. *The Walking Cure: Pep And Power from Walking; How to Cure Disease by Walking.* New York: Macfadden Publications, 1925.

Macfadden, Bernarr. *Womanhood and Marriage.* New York: Physical Culture Publishing Co., 1918.

Macfadden, Johnnie Lee. *Barefoot in Eden: The Macfadden Plan for Health, Charm and Long-Lasting Youth.* Englewood Cliffs, NJ: Prentice-Hall, 1962.

Macfadden, Mary, with Emile Gauvreau. *Dumbbells and Carrot Strips: The Story of Bernarr Macfadden.* New York: Henry Holt, 1953.

Maguire, James. *Impresario: The Life and Times of Ed Sullivan.* New York: Billboard Books, 2006.

Mallen, Frank. *Sauce for the Gander.* White Plains, NY: Baldwin Books, 1954.

Manchester, Harland. "True Stories." *Scribner's Magazine,* Aug. 1938.

McNulty, John Bard. *Older than the Nation: The Life and Times of the Hartford Courant.* Stonington, CT: Pequot Press, 1964.

Morgan, Thomas B. *Italian Physical Culture Demonstration: A Report of the Visit, Training and Accomplishments of the Forty Italian Students Who Were Guests of Bernarr Macfadden During a Stay of Six Months in the United States Studying His Methods of Physical Culture.* New York: Macfadden Book Co., 1932.

Morris, Edmund. *The Rise of Theodore Roosevelt.* New York: Random House, 1979.

Mowbray, Scott. *The Food Fight: Truth, Myth and the Food-Health Connection.* Toronto: Random House, 1992.

Mrozek, Donald J. *Sport and American Mentality, 1880–1910.* Knoxville: University of Tennessee Press, 1983.

Nagel, Paul C. *Missouri, A Bicentennial History.* New York: W. W. Norton, 1977.

Nasaw, David. *The Chief: The Life of William Randolph Hearst.* New York: Houghton Mifflin, 2000.

Oursler, Fulton, with Fulton Oursler Jr. *Behold This Dreamer! An Autobiography.* Boston: Little, Brown, 1964.

Oursler, Fulton. *The True Story of Bernarr Macfadden.* New York: Lewis Copeland, 1929.

Oursler, Grace Perkins. *Chats with the Macfadden Family.* New York: Lewis Copeland, 1929.

Oursler, Will. *Family Story.* New York: Funk & Wagnalls, 1963.

Peterson, Theodore. *Magazines in the Twentieth Century,* 2d ed. Urbana: University of Illinois Press, 1964.

Rodale, J. I. *Pay Dirt: Farming and Gardening with Composts.* New York: Devin-Adair, 1945.

Rosen, George. *A History of Public Health,* expanded ed. Baltimore: Johns Hopkins University Press, 1993.

Sampson, Robert. *Yesterday's Faces: A Study of Series Characters in the Early Pulp Magazines.* Bowling Green, OH: Bowling Green State University Popular Press, 1983.

Schwartz, Hillel. *Never Satisfied: A Cultural History of Diets, Fantasies, and Fat.* New York: Free Press, 1986.

Sinclair, Upton. *American Outpost: A Book of Reminiscences.* New York: Farrar & Rinehart, 1932.

Sinclair, Upton. *Autobiography of Upton Sinclair.* New York: Harcourt, Brace & World, 1962.

Sinclair, Upton. *The Fasting Cure.* Pasadena, CA: privately published, 1911.

Taft, William H. "Bernarr Macfadden." *Missouri Historical Review,* Oct. 1968.

Taylor, Robert Lewis. "I Was Once a 97-Pound Weakling." *New Yorker,* Jan. 3, 1942.

Taylor, Robert Lewis. "Physical Culture," parts 1, 2, and 3. *New Yorker,* Oct. 14, 21, and 28, 1950.

Tebeau, Charlton. *History of Florida.* Coral Gables, FL: University of Miami Press, 1971.

Tilney, Frederick. *Young at 73—and Beyond!* New York: Information, Inc., 1968.

Todd, Jan. "Bernarr Macfadden: Reformer of Feminine Form," *Journal of Sport History,* Spring 1987.

Todd, Jan. *Physical Culture and the Body Beautiful: Purposive Exercise in the Lives of American Women, 1800–1870.* Macon, GA: Mercer University Press, 1998.

Toomer, Jean. *The Wayward and the Seeking: A Collection of Writings by Jean Toomer,* ed. Darwin T. Turner. Washington, DC: Howard University Press, 1980.

Trager, James. *The New York Chronology.* New York: HarperCollins, 2003.

Vanderbilt, Cornelius. *Man of the World: My Life on Five Continents.* New York: Crown, 1959.

Walker, Stanley. *Mrs. Astor's Horse.* New York: Frederick A. Stokes, 1935.

Waugh, Clifford. *Bernarr Macfadden: The Muscular Prophet.* PhD dissertation, University of Buffalo, 1979.

Weider, Joe, Ben Weider, and Mike Steere. *Brothers of Iron*. Champaign, IL: Sports Publishing, 2006.

Whorton, James C. *Crusaders for Fitness: The History of American Health Reformers*. Princeton, NJ: Princeton University Press, 1982.

Whorton, James C. *Nature Cures: The History of Alternative Medicine in America*. New York: Oxford University Press, 2002.

Wood, Clement. *Bernarr Macfadden: A Study in Success*. New York: Lewis Copeland, 1929.

Young, James Harvey. *The Medical Messiahs: A Social History of Health Quackery in Twentieth-Century America*. Princeton, NJ: Princeton University Press, 1967.

Zuckerman, Mary Ellen. *A History of Popular Women's Magazines in the United States 1792–1995*. Westport, CT: Greenwood Press, 1998.

INDEX